Advances in Intelligent Systems and Computing

Volume 272

Series editor

Janusz Kacprzyk, Polish Academy of Sciences, Warsaw, Poland
e-mail: kacprzyk@ibspan.waw.pl

For further volumes:
http://www.springer.com/series/11156

About this Series

The series "Advances in Intelligent Systems and Computing" contains publications on theory, applications, and design methods of Intelligent Systems and Intelligent Computing. Virtually all disciplines such as engineering, natural sciences, computer and information science, ICT, economics, business, e-commerce, environment, healthcare, life science are covered. The list of topics spans all the areas of modern intelligent systems and computing.

The publications within "Advances in Intelligent Systems and Computing" are primarily textbooks and proceedings of important conferences, symposia and congresses. They cover significant recent developments in the field, both of a foundational and applicable character. An important characteristic feature of the series is the short publication time and world-wide distribution. This permits a rapid and broad dissemination of research results.

Sungshin Kim · Jin-Woo Jung
Naoyuki Kubota
Editors

Soft Computing in Intelligent Control

Springer

Editors
Sungshin Kim
Pusan National University
Busan
Korea

Naoyuki Kubota
Tokyo Metropolitan University
Tokyo
Japan

Jin-Woo Jung
Dongguk University
Seoul
Korea

ISSN 2194-5357 ISSN 2194-5365 (electronic)
ISBN 978-3-319-05569-5 ISBN 978-3-319-05570-1 (eBook)
DOI 10.1007/978-3-319-05570-1
Springer Cham Heidelberg New York Dordrecht London

Library of Congress Control Number: 2014933550

Printed on acid-free paper

Springer is part of Springer Science+Business Media (www.springer.com)

Preface

Nowadays, people have tendancy to be fond of smarter machines that are able to collect data, make learning, recognize things, infer meanings, communicate with human and perform behaviors. Thus, we have built advanced intelligent control affecting all around societies; automotive, rail, aerospace, defense, energy, healthcare, telecoms and consumer electronics, finance, urbanization. Consequently, users and consumers can take new experiences through the intelligent control systems. We can reshape the technology world and provide new opportunities for industry and business, by offering cost-effective, sustainable and innovative business models. We will have to know how to create our own digital life. The intelligent control systems enable people to make complex applications, to implement system integration and to meet society's demand for safety and security. This book aims at presenting the research results and solutions of applications in relevance with intelligent control systems. We propose to researchers and practitioners some methods to advance the intelligent controls and apply the intelligent control to specific or general purpose. This book consists of 10 contributions that feature an experimental verification of defect detections, depth-based visual object groupings, fuzzy-tuning PID controller, and control of traffic speed, robust object detection, and detection method of radio frequency interference, ontological model for the tax system, future toy web, cooperation level estimation, and interface for wearable computers. This edition is published in original, peer reviewed contributions covering from initial design to final prototypes and authorization.

To help readers understand articles, we describe the short introduction of each article as follows;

1. "Experimental verification and analysis for ability of defect detection with angle for axis direction of guided wave using magnetostrictive sensor": This paper presents an experiment of transmitting longitudinal mode and torisional mode of guided wave using magnetostrictive sensor for test plate of mock- up with artificial defect of 90°, 75° 60° for axis direction of pipe. Each characteristic of each mode compares with Vpp along with angle for axis direction of received defect signal.

2. "System for Proving N-screen Service using Depth-based Visual Object Groupings": This article is about the system development which it groups the independent visual

objects with the group value unit and extracts each grouping objects with the interesting object unit and interaction with the user in the various screen devices.
3. "Fuzzy-Tuning PID Controller for Nonlinear Electromagnetic Levitation System": The paper derives a dynamic model of the magnetic levitation system and proposes a Fuzzy-Tuning PID (FTP) controller that selects the parameters of the PID controller by using fuzzy inference system. Conventional PID controller can be applied to control the electromagnet levitation.
4. "An Effect of Traffic Speed on Maritime Accidents": This paper provides persuasive evidence to enforce a safe speed to reduce ship collisions in VTS area. The authors have investigated the correlation among statistical parameters of vessel speed, traffic flow, and potential collision rate in VTS area.
5. "Robust Object Detection in Sea Environment Based on DWT": This paper presents an approach to detecting object regions considering horizontal and vertical edges. Firstly, image enhancement is performed by suppressing noises such as sea glint and complex clutters using a statistical filter. Secondly, a horizontal edge map and a vertical edge map are generated by Discrete Wavelet Transform. Thirdly, a saliency map integrating the horizontal and the vertical edge maps is generated. Finally, object regions are detected by an adaptive thresholding method.
6. "Detection Method of Radio Frequency Interference Using Raw Measurement of Multi-GNSS Receivers": This article proposes the simple method for detecting the RFI at current DGNSS station, and shows the test results by using multi-DGNSS receivers. It introduces the configuration and functionalities of DGNSS RSIM, and presents the limitation of integrity monitoring function, when the RFI occurred in the current DGNSS station.
7. "Creating an ontological model for the tax system in Kazakhstan": In this paper, an ontological model will be created to provide corresponding operating conditions of the system and implement a web-application that uses Semantic Web technologies and represents the interface of the expert system machine inference that works with these concepts.
8. "2030 Toy Web of the Future": This paper defines the toy web as web-based interactive contents for children and teen agers to improve their cognitive ability. Its market is just emerging now and companies anticipate a foresight how its user interface in 2030 will look like. In order to foresight we have to identify key drivers and related variables.
9. "Cooperation Level Estimation of Pair Work Using Top-view Image": This paper proposes a method for estimating a cooperation level in pair work. The task is a cooperation work that takes place in front of a whiteboard by two persons. It is applicable to two machine learning approaches: AdaBooost and multiple linear regressions.
10. "Korean Word Search Interface for Wearable Computers Using a Wrist-Mounted Camera Device": This paper presents a hand shape recognition method and its application to Korean word search interface for wearable computers. A user adaptation process is required for algorithmic details and recognition performance.

We would appreciate it if readers could get useful information from the articles and contribute to creating innovative and novel concept or theory. Thank you.

Sungshin Kim
Jin-Woo Jung
Naoyuki Kubota

Contents

Experimental Verification and Analysis for Ability of Defect Detection with Angle for Axis Direction of Guided Wave Using Magnetostrictive Sensor

Yigon Kim[1], Youngchul Bae[1], Chen Li[1], Jeong-Ki Lee[2], Hongsik Moon[2], Deahwa Yoo[2], and Kyung-Jo Park[3]

[1] Division of Electrical, Electronic and Computer Eng., Chonnam National University
50 Daehak-ro, Yeosu, Jeonnam, 550-749, Korea
{yigon,ycbae}@chonnam.ac.kr, cl352227@gmail.com
[2] Center for Safety Diagnosis Technology of Heavy & Chemical Facilities, Chonnam National University
50 Daehak-ro, Yeosu, Jeonnam, 550-749, Korea
{leejki,mhs99co,fantable}@chonnam.ac.kr
[3] Department of Mechanical Engineering, Chonnam National University
50 Daehak-ro, Yeosu, Jeonnam, 550-749, Korea
kjpark40@chonnam.ac.kr

Abstract. In this paper we induced it by using disperse which is wave analysis program developed by Imperial College of United Kingdom. In this paper we sent and received longitudinal mode and torisional mode of guided wave using magnetostrictive sensor for test plate of mock-up with artificial defect of 90°, 75° 60° for axis direction of pipe. We review each characteristics of each mode compare with Vpp along with angle for axis direction of received defect signal.

1 Introduction

The most of petrochemical industrial complex in global have been used many pipes as means of transport of main raw material and product. In order to secure safety of pipe and take precautions against possible accidents, the pipes have been performed regular inspection with various non-destructive inspections. The regular tests have performed not only examination with the naked eye but also local sightseeing such as RT(Radiography Test), UT(Ultrasonic test),PT(Penetrant test) and ECT(eddy current test). However, such a local sightseeing spends many time and cost to inspect whole systems. Therefore, many researches are progressing for pipe inspection using guided wave as a method that find a serious pipe's defect during short time for whole system in the beginning of inspection [1-5].

That guided wave that could use for inspection of pipe may largely classify in three types such as longitudinal mode, flexural mode and torsional mode. The flexural mode never used to test guided wave because the mode shape has an asymmetry. Recently the pipe inspection through sending and receiving of torisional mode using magnetostrictive and piezoelelctric sensor has been largely interested [6-9].

S. Kim et al. (eds.), *Soft Computing in Intelligent Control*,
Advances in Intelligent Systems and Computing 272,
DOI: 10.1007/978-3-319-05570-1_1, © Springer International Publishing Switzerland 2014

The guided wave have a disperse characteristics that are varied according to frequency, and it was known that is able to occur distortion of wave as a travelling wave along with wave guide. In order to analyze dispersion characteristics and distortion of wave we need analysis of dispersion plot of wave guide.

In this paper we induced it by using Disperse which is wave analysis program developed by Imperial College of United Kingdom. In this paper we send and received longitudinal mode and torisional mode of guided wave using magnetostrictive sensor for test plate of mock-up with artificial defect of 90°, 75° 60° for axis direction of pipe. We review each characteristics of each mode compare with Vpp along with angle for axis direction of received defect signal.

2 Manufacturing of Mock-Up Test Plate

We make the test plate of mock-up with manufacturing artificial defect to the steel pipe in order to verify experimentally guided wave for pipeline of petrochemical industrial complex. The mock-up test plate made by specification of KSD 3652 SPPS380 as steel pipe for pipe arrangement of pressure; it has 114.3mm of external diameter, 6.02mm (SCH 40) of thickness, and 1980mm of total length. Fig. 1 shows the artificial defect manufactured at pipe with 60, 77 and 90 degree, respectively.

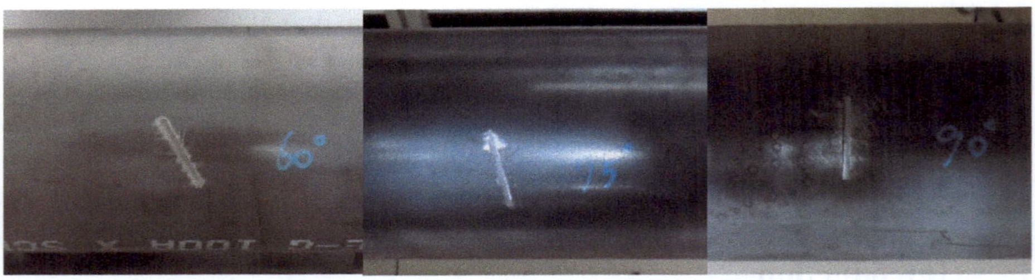

Fig. 1. Artifical defect manufactured at pipe (60, 75, 90 degree from left)

3 Plot of Dispersion Velocity

The propagation characteristics of guided wave displayed at pipe have a very complicate aspect. The graph of group velocity of 0~250KHz for target pipe shows in Fig. 2.

There are wavelengths of n unit (n=0,1,2,3,••) along with circumference in structures of shell shape. If n is equal to 0 then wavelength had an axis symmetry mode, and n is greater than 1 then the propagating wavelength had a circumference direction and axis direction. The mode was described by T(n,m), L(n,m), F(n,m) like description method of Silk and Bainton [10]. Where n,m is integer, and T,L,F represent longitudinal mode, flexural mode and torsional mode.

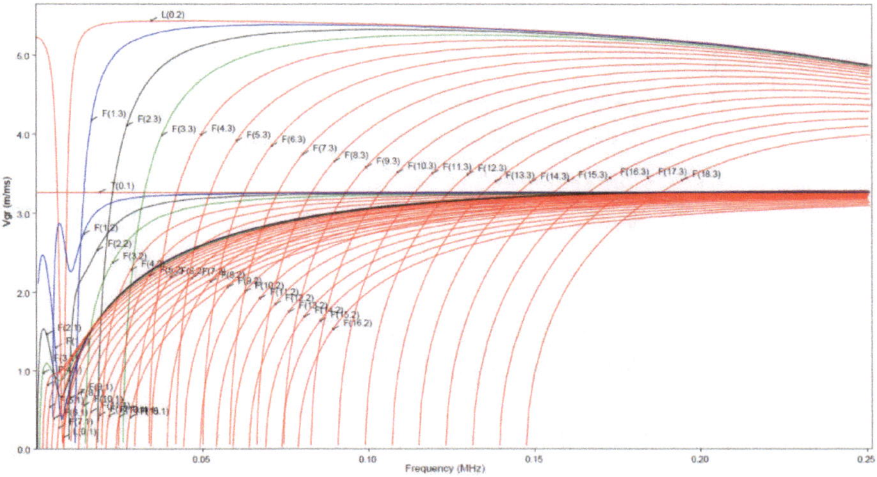

Fig. 2. Group velocity dispersion curves for the pipe under inspection)

In general ultrasonic sightseeing, speed is constant along with the quality of material without relation of frequency. However guided wave has a characteristic that vary speed according to the quality of material and frequency. Hence, we have to find applying frequency domain with constant speed when we set testing frequency. This frequency is called “0 “as regime.

In this paper, we send and receive the T(0,1) mode and L(0,2) mode and we eliminate frequency domain of less than 25kHz that L(0,2) mode represents dispersion characteristics in order to detect artificial defect manufactured at test plate of mock-up. The length of test plate is 1980 mm. Since it is shorter wavelength relatively, we decided to use more than 100 kHz frequency bandwidth that has excellent detection ability and has relatively short of bandwidth of wavelength.

4 Sending and Receiving of Longitudinal Mode and Torisional Mode by Magnetostrictive Sensor

The torisional mode was send and receives by crossed-coil magnetostrictive sensor. [11]. The way of sending and receiving of torisional mode are reported by pre-magnetization[12], method of use by ferromagnetic patch and ferromagnetic patch with 45 degree, and method of use by crossed-coils sensor. In this paper we sending and receiving torisional mode by using crossed-coils sensor that are able to sending for certain mode (T(0.1)) and have S/N ratio. Fig. 3 shows structure of crossed-coils sensors.

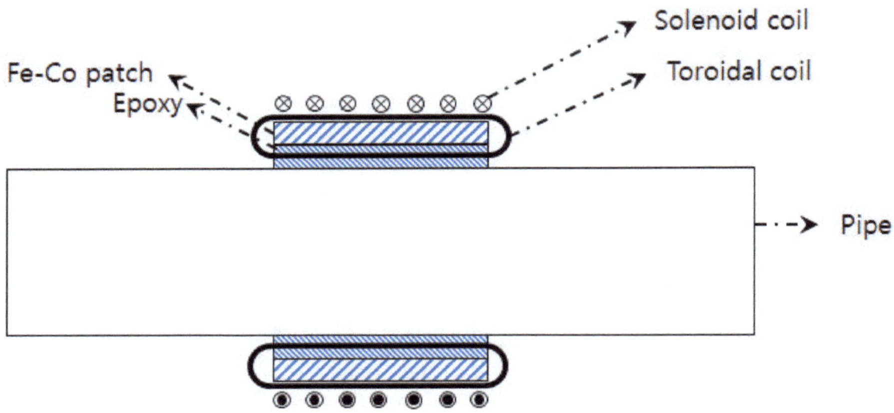

Fig. 3. Sending and receiving of torisional mode

As sending and receiving method of longitudinal mode, we fixed ferromagnetic patch (Fe-co) to the pipe by epoxy etc. We also manufactured second solenoid coil (1300 turn) and then DC bias of 2.5A was flown to the pipe to create time in-varying magnetic field of an axis direction. Fig. 4 shows a method of sending and receiving of longitudinal mode using bias magnetic field.

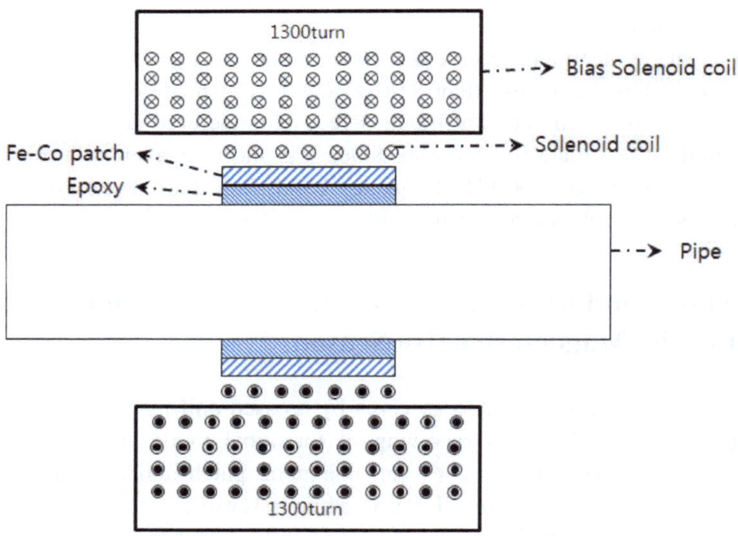

Fig. 4. Sending and receiving of longitudinal mode

5 Experimental Result

5.1 Sending and Receiving of Torisional Mode

We represent result of sending and receiving of T(0,1) mode of guided wave at test plate of artificial defect with 60°, 75°, 90° for axis direction through Fig. 5 - Fig. 7. Table 1 represents magnitude variation of defect signal according to decrees of angular for axis direction.

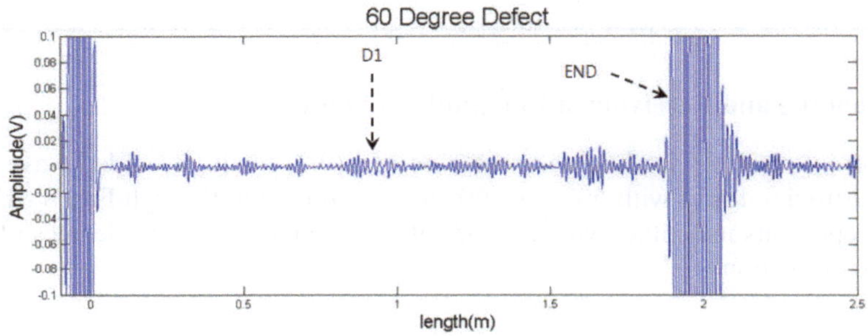

Fig. 5. Receiving signal of torisional mode with 60 degree

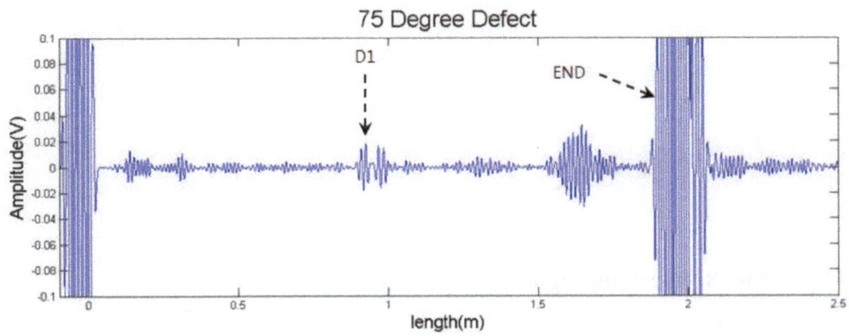

Fig. 6. Receiving signal of torisional mode with 75 degree

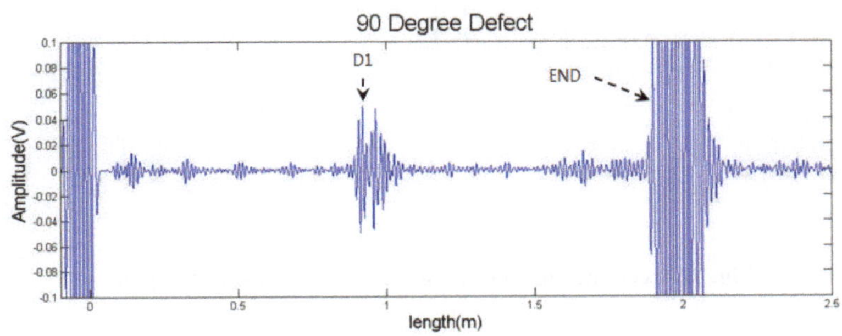

Fig. 7. Receiving signal of torisional mode with 90 degree

Table 1. Magnitude of receiving signal for torisional mode with each angle for artificial defects

Angular of artificial defect for axis direction	Vpp(V) of received signal at defect
60°	0.01578
75°	0.03574
90°	0.09683

5.2 Sending and Receiving of Longitudinal Mode

We represent result of sending and receiving of L(0,2) mode of guided wave at test plate of artificial defect with 60°, 75°, 90° for axis direction through Fig. 8 - Fig. 10. Table 2 represents magnitude variation of defect signal according to decrees of angular for axis direction.

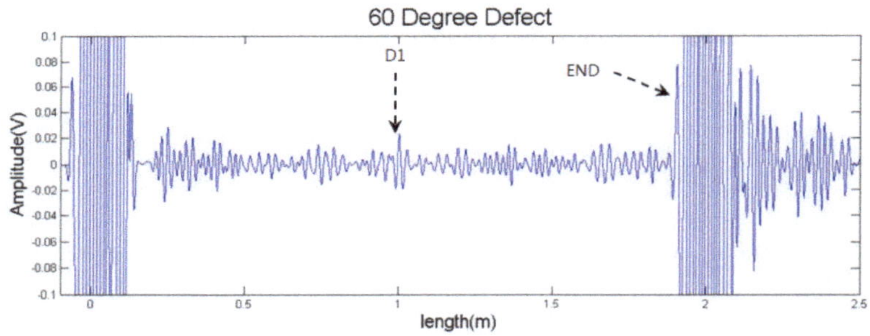

Fig. 8. Receiving signal of longitudinal mode with 60 degree

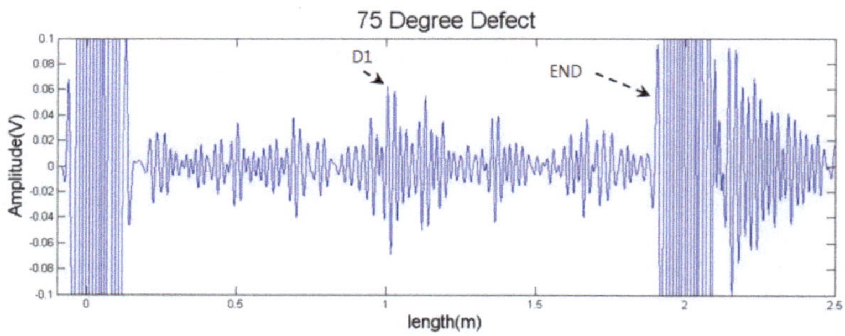

Fig. 9. Receiving signal of longitudinal mode with 75 degree

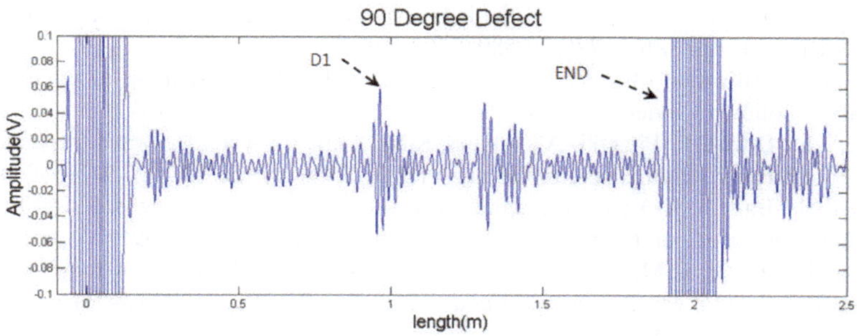

Fig. 10. Receiving signal of longitudinal mode with 90 degree

Table 2. Magnitude of receiving signal for longitudinal mode with each angle for artificial defects

Angular of artificial defect for axis direction	Vpp(V) of received signal at defect
60°	0.04118
75°	0.13074
90°	0.11283

6 Result

From our experiment we recognize the variation of signal magnitude according to variation of defect angle of axis direction of each mode. The more decrease an angle of defect, the returned signal will decrease more, which reflected from defect part both in torsional mode (T0.1) and in longitudinal mode (L0.2). At longitudinal mode (L0.2), we recognize that the defect signal of 90 degree is more large compare to 75 degree, we judge that this signal contains induce ultrasonic signals. We also know that reduction rate of defect signal according to angle of defect in longitudinal mode (L0.2) is low compare to torsional mode (T0.1).

Acknowledgment. This study was supported by the Ministry of Knowledge Economy (MKE) through the Regional Innovation Centre Programme.

References

1. Brook, M., Ngoc, T.D.K., Eder, J.: Ultrasonic inspection of steam generator tubing by cylindrical guided wave. In: Review of Progress in Quantitative Nondestructive Evaluation, vol. 9, pp. 243–249 (1990)

2. Ditri, J.J., Rose, J.L.: Excitation of guided wave modes in hollow cylinders by applied surface traction. J. App. Phys. 72(7), 2589–2597 (1992)
3. Ditri, J.J.: Utilization of guided elastic waves for the characterization of circumferential cracks in hollow cylinders. J. Acoust. Soc. Am. 96, 3769–3775 (1994)
4. Rose, J.L., Ditri, J.J., Pilarski, A., Rajana, K.M., Carr, F.: A guided wave inspection technique for nuclear steam generator tubing. NDT & E International 27(6), 307–310 (1994)
5. Kwun, H., Bartels, K.A.: Experimental observation of elastic wave dispersion in bonded solids various configurations. J. Acoust. Soc. Am. 99(2), 962–968 (1996)
6. Kwun, H., Teller, C.M.: Magnetostrictive generation and detection of longitudinal, torsional, and flexural waves in rod. J. Acoust. Soc. Am. 96, 1202–1204 (1994)
7. Kwun, H., Batels, K.A.: Experimental observation of wave dispersion in cylindrical shells via time-frequency analysis. J. Acoust. Soc. Am. 97(6), 3905–3907 (1995)
8. Kwun, H., Batels, K.A.: Magnetostrictive sensor technology and its application. Ultrasonic 36, 171–178 (1998)
9. Lee, H.C., Kim, Y.Y.: Wave selection using a magnetomechanical sensor in a solid cylinder. J. Acoust. Soc. Am. 112(3), 953–960 (2002)
10. Silk, M.G., Bainton, K.F.: The propagation in metal tubing of ultrasonic mode equivalent to Lamb waves. Ultrasonics 17(1), 9–11 (1979)
11. Kim, Y.G., Moon, H.S., Park, K.J., Lee, J.K.: Generating and detecting torsional guided waves using magnetostrictive sensors of crossed coils. NDT & E International 44(2), 145–151 (2011)
12. Kwun, H.: Method and apparatus generating and detevtingtorional wave inspection of pipe or tubes. United States Patent, Patent No: US6429650 B1 (2002)

System for Proving N-screen Service Using Depth-Based Visual Object Groupings

Kwang-Yong Kim, Il-Gu Jung, and Won Ryu

Intelligent Convergence Media Research Department, ETRI 218 Gajeongno,
Yuseong-gu, Daejeon, Korea
{kwangyk,ilkoo,wlyu}@etri.re.kr

Abstract. This paper is the thing about the system development which it groups the independent visual objects with the group value unit and extracts each grouping objects with the interesting object unit and can interaction with the user in the various screen devices.

1 Introduction

Today, when watching video, if visual objects are overlapped with each other, we cannot (always) clearly see an object that we wish to see because of occluding objects. Also, we cannot show visual objects that are hidden at time that want while hiding other objects to occlusion region intentionally. Specially, in the case of 3D video service, we cannot watch again these 2D visual objects after separating 2D visual objects that we are interested according to sight depth degree of 3D visual object that are watching. Furthermore, we can get the service effect of superior targeting advertising if we watch to each interesting object in own screen in each user by combining again these objects after separating interesting objects. In this paper from now, we try to define an object that wish to watch as the term called interesting object. Now, we classify into the group objects which have the same depth according to their depth degree about perfectly unseen visual objects by being occluded each other. Thus, we designed the interactive N-screen system that can watch each object of interest and group objects in own screen in each user by providing the function of the event handling so that users can do interaction with group objects

2 The Trends of Related Research

2.1 The Research Trends on the Object Segmentation

In general, segmentation can be defined as the process of partitioning data into groups of potential subsets that share similar characteristics. It has become a key technique for semantic content extraction and plays an important role in digital multimedia processing, pattern recognition, and computer vision. The task of image/video segmentation emerges in many application areas, such as image interpretation, video analysis and understanding, video summarization and indexing, and digital

S. Kim et al. (eds.), *Soft Computing in Intelligent Control*,
Advances in Intelligent Systems and Computing 272,
DOI: 10.1007/978-3-319-05570-1_2, © Springer International Publishing Switzerland 2014

entertainment [1]. Segmentation is also used to partition each frame of a video sequence into semantically meaningful objects with arbitrary shape. In the past several years, the problem of segmenting image/video data has become a fundamental one and had significant impact on both new pattern recognition algorithms and applications [1]. The goal of image segmentation is very application-oriented, which emerges in many fields. A limited set of applications of image/video segmentation can be presented as follows: Object recognition, where the segmentation is treated as a key component that groups coherent image areas that are then used to assemble and detect objects. As important recognition tasks, feature extraction and model matching rely heavily on the quality of the image segmentation process. When an image is segmented into several homogeneous intensity regions, each region can be used as features for deriving the category model since they are rich descriptors, usually stable to small illumination and viewpoint changes. Video monitoring, where an object can be divided into pieces to improve tracking robustness to occlusion by tracking the evolution of the moving objects along the time axis[1] associated with the segments. An ordered list of segments associated with the query object will be returned to user, which has been applied to the content classification, representation, or understanding .Data compression, which allows suitable coding algorithm to manipulate each object independently resulting in subjective quality improvement [1]. The MPEG-4 standard supports the object-based coding of a video sequence, i.e., video objects, the partitioning of the video sequence into spatial-temporal regions that correspond to meaningful objects within the video sequence [3]. They implemented the system which Voronoi Order and a surface optimization problem formulation is combined for automatic video objects segmentation as the major enabling technology for this standard in this paper[3]. The segmentation method is largely classified into region-based segmentation, data clustering, and edge-base segmentation [4]. Region-based methods mainly rely on the assumption that the neighboring pixels within one region have similar value [4-5]. The common procedure is to compare one pixel with its neighbors. If a similarity criterion is satisfied, the pixel can be set belong to the cluster as one or more of its neighbors. the seeded region growing, the Unseeded region growing, the Region splitting and merging, and the Fast scanning algorithm belongs to the Region-based methods, where the seeded region growing (SRG) algorithm performs a segmentation of an image with examine the neighboring pixels of a set of points, known as seed points, and determine whether the pixels could be classified to the cluster of seed point or not. The unseeded region growing (URG) algorithm is a derivative of seeded region growing and no explicit seed selection is necessary. So this method can perform fully automatic segmentation with the added benefit of robustness from being a region-based segmentation. Region splitting and merging is to distinguish the homogeneity of the image that is based on quad-trees, which it means each node of trees has four descendants and the root of the tree corresponds to the entire image. Besides, each node represents the subdivision of a node into four descendant nodes. Next, fast scanning algorithm is to scan from the upper-left corner to lower-right corner of the whole image and to decide if the examined pixel can merged into an existed clustering, where the merged criterion is based on our assigned threshold, where if the difference between the pixel value and the average pixel value of the adjacent cluster is smaller than the threshold, then this pixel can be merged into the cluster. The basic concept of data clustering is to use the centered to represent each cluster and base on the similarity with the centered of cluster to classify. We can roughly classify into the hierarchical clustering, the partitioned clustering and the mean shift algorithm basing on density estimation. Edge-base segmentation methods apply some edge detection methods before segmentation, for instance, watershed segmentation algorithm [3]. Now, what is the most important point that different from the static image segmentation and video object segmentation? It lies in the usage to motion information. The main concept of video object segmentation chiefly makes use of motion information and it is that separates the foreground object from its background by combining with information about other colors, textures, edges, regions and so on[5]. The following briefly introduces some common methods [5]. There is the video segmentation technology based on the temporal information. It is the identification method of motion object in which it includes change detection, optical flow method, object tracking, motion estimation. The basic concept of change detection is that the current video frame is segmented into changed area and unchanged area relative to reference frame through the detection of frame difference between the former and latter frame, unchanged part represents static background, changed part represents motion and occlusion [5-7]. Another one is method called optical flow. When human watches movement objects, object image forms a series of consecutive change images on the retina, the series of change information is continuously "flowing" the retina, like a light "flow". The optical flow method is a common method used to estimate the motion field. But there is the occlusion/exposed problem for the method and also steady assumption of optical flow field can not be satisfied at the point of the brightness mutation. So far, we have looked at the video segmentation technology based on the temporal information. Another technique is the video segmentation technology based on the spatial-temporal information. Segmentation method of spatial-temporal joint is based on spatial and temporal clustering segmentation method, based on the tracking or judgment method of region theory, based on object tracking method and so on[5]. But, these pixel based Segmentation technologies require time-consuming pixel-based operations such as watershed algorithm, region growing, and dense motion field estimation to obtain spatial and temporal information from each pixel in the frames. So, they usually take several seconds to find or track an object in a frame [5-6]. In [8], the authors proposed a fast video object segmentation algorithm that extracts objects from H.264 bit stream. This utilizes the motion information from the H.264 compressed bit stream to identify background motion model and moving objects. In order to preserve spatial and temporal continuity of objects, Markov random field (MRF) is used to model the foreground field. Quantized transform coefficients of the residual frame are also used to improve segmentation result.

2.2 The Research Trends on Video Object Tracking

The aim of an object tracking is to generate the trajectory of an object over time by locating its position in every frame of the video [2]. Tracking is usually performed in the context of higher-level applications that require the location and/or shape of the object in every frame [2]. In [2], the authors were classified into five tracking methods as follows; In the first case, it is the point tracking method, which it is Objects detected in consecutive frames are represented by points, and a point matching is done. This approach requires an external mechanism to detect the objects in every frame. In the second case, it is the kernel tracking method where kernel refers to the object shape and appearance. Kernel can be a rectangular template or an elliptical shape with an associate histogram. Objects are tracked by computing the motion of the kernel in consecutive frames. This motion means the form of a parametric transformation such as translation, rotation, and affine. In the last case, it is the silhouette tracking method where it is performed by estimating the object region in each frame by using the information encoded inside the object region. This information can be in the form of appearance density and shape models which are in the form of edge maps. Given the object models, silhouettes are tracked by either shape matching or contour evolution which can be considered as object segmentation applied in the temporal domain using the priors generated from the previous frames.

3 The Design of Interactive N Screen Service System Using Group Visual Objects Based on Depth

When we watch interesting objects which are occluded with other objects on TV, we can't see completely independent interesting objects.

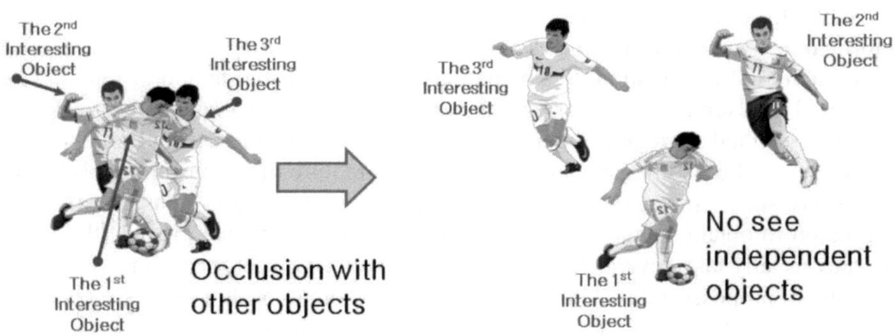

Fig. 1. The case of service which can't watch completely independent interesting objects

Furthermore, as shown in Figure 1, after service providers hid interesting objects at the region which are occluded intentionally, when interesting objects are extracted, they cannot provide the service to see hidden objects. Figure 2 shows an example of service which creates intentionally hidden object among occluded objects and provides an watcher with them using N-screen devices.

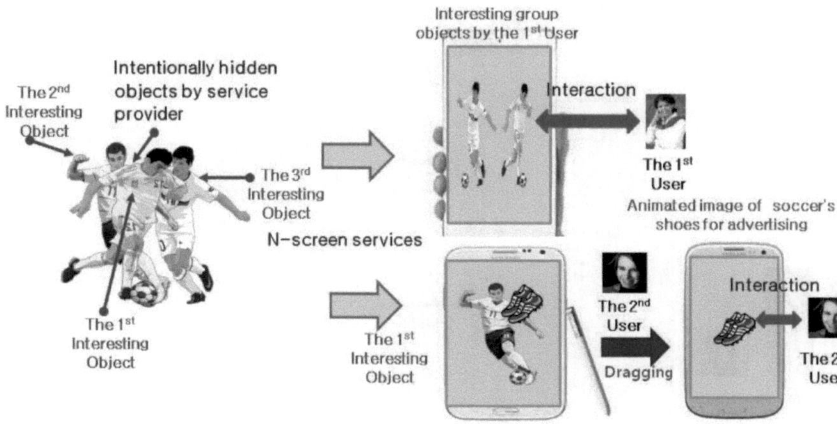

Fig. 2. The service which creates purposely hidden object among occluded objects and provides an watcher with them using N-screen devices

3.1 The Design of Interactive N Screen Service System Using Group Visual Objects Based on Depth

In this paper, we classify into the group objects which have the same depth according to their depth degree about perfectly unseen visual objects by being occluded each other. Thus, we designed the interactive N-screen system that can watch each object of interest and group objects in own screen in each user by providing the function of the event handling so that users can do interaction with group objects. Figure 3 shows the architecture of authoring module which can edit group object by group unit or edit each of them. Figure 4 shows the architecture of authoring module which can edit group object by group unit or edit each of them. As shown in Figure 3 and Figure 4, the first stage is which we extract visual objects automatically or semi-automatically by using the Auto/Semi-auto Visual object Extractor and is which stored independent visual object files in the D/B. The second stage is the Authoring Stage of Group and Independent Object, where we can edit and modify for spatial/temporal playing & event handling of independent visual objects. When executing the authoring for spatial/temporal playing & event handling of group object having the same depth value, we create group objects of same depth as the number of N screen for rendering. Third stage is to create and manage the group & Independent visual scene tree, where we create group & independent visual object tree according to the structural rules of scene tree, object descriptor, and event handling for Interaction. We also create associated information of object and these nodes are stored by the structure of XML DOM tree. Furthermore, we process the creation stage of MP4 file and finally execute the streaming stage of MP4 file as shown by Figure 3 and Figure 4.

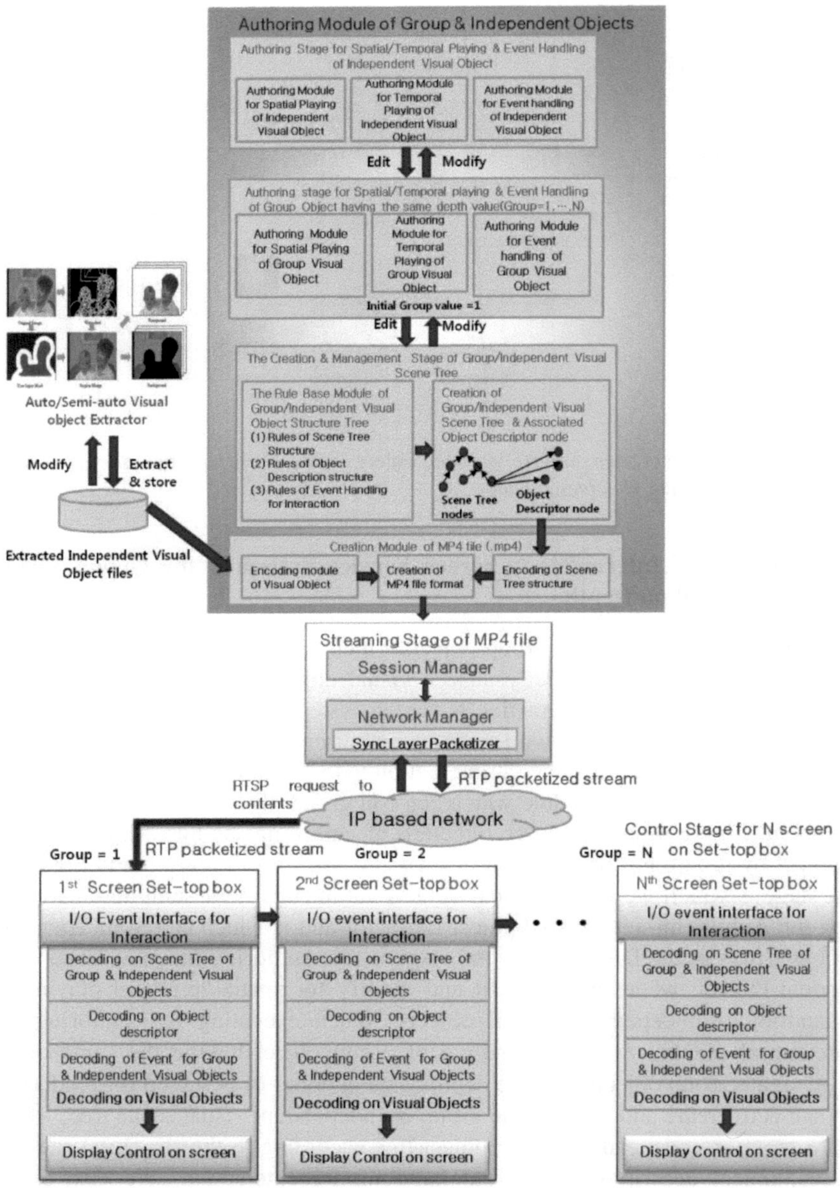

Fig. 3. The architecture of authoring module which can edit by group unit or edit each of objects

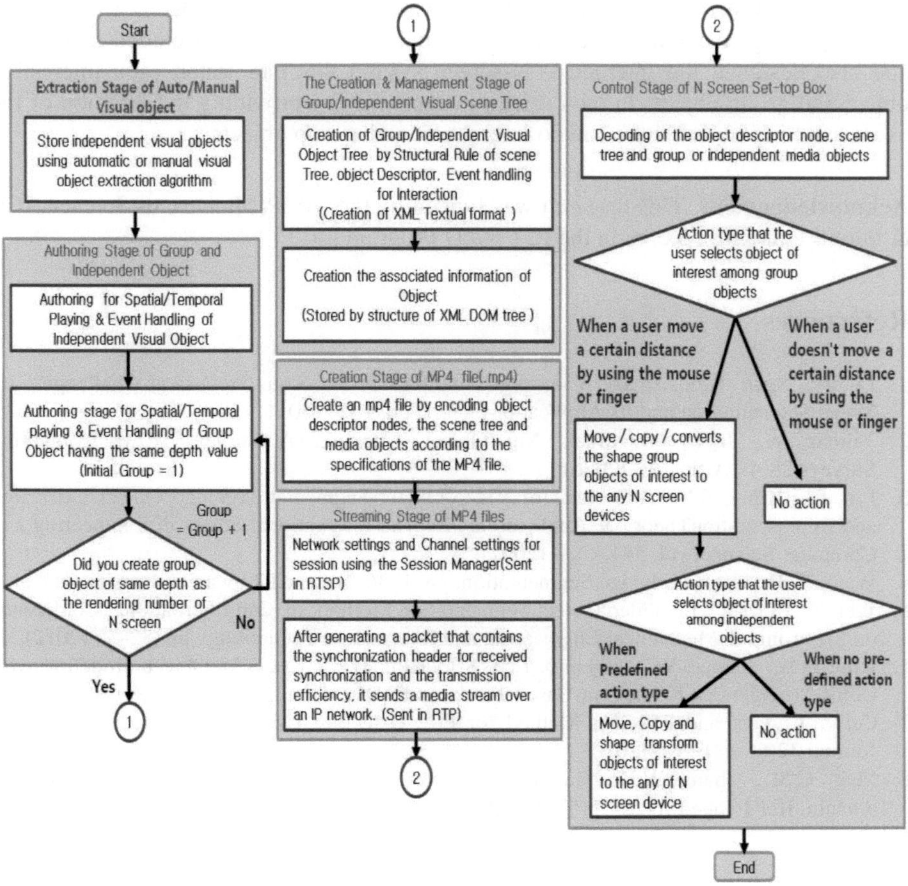

Fig. 4. Block Diagram of authoring module which can edit by group unit or edit each of objects

4 Conclusions

Nowadays, if visual objects are overlapped with each other when watching the video, we cannot see clearly objects that wish to see because of objects that is overlapped. Besides, we cannot show visual objects that are hidden at time that want when hiding other objects to occlusion region purposely. Furthermore, in the case of 3D video service, we cannot watch again these 2D visual objects after separating 2D visual objects that we are interested according to sight depth degree of 3D visual object that are watching. In addition, we can get the effects of superior targeting advertisement service if we watch to each interesting object in own screen in each user by combining again these objects after separating interesting objects by means of the depth degree of visual objects shown in each multi screen used in the service of digital signage. Therefore, in order to solve these issues, we designed the system which can classify into the group objects which have the same depth according to their depth degree about perfectly unseen visual objects by being occluded each other, and also designed the interactive N-screen system that can watch each interesting object and group objects in own screen in each user by providing the function of the event handling so that users can do interaction with group objects.

Acknowledgement. This research was funded by the MSIP(Ministry of Science, ICT & Future Planning), Korea in the ICT R&D Program 2013.

References

1. Li., H., Ngan, K.N.: Image Videosegmentation Current StatusTrendsand Challenges. In: Video Segmentation and Its Applications, ch. 1, pp. 1–23 (2011)
2. Yilmaz, A., Javed, O., Shah, M.: Object Tracking: A Survey. ACM Computing Surveys 38(4), Article 13 (2006)
3. Lin, I.J., Kung, S.Y.: A System for Video Object Segmentation-Video Object Extraction and RepresentationTheory & Applications. In: The International Series in Engineering and Computer Science, vol. 584, ch. 4 (2002)
4. Wang, Y.H.: Tutorial: Image Segmentation, pp. 1–36. National Taiwan University, Taipei
5. Wei, Z.: Research on Video Object Segmentation Method. In: 2nd International Conference on Electronic & Mechanical Engineering and Information Technology, pp. 52–56 (2012)
6. Kim, D.H., Jung, S.W., Suryanto, Lee, S.J., Kim, H.K., Ko, S.J.: Object Modeling with Color Arrangement for Region-Based Tracking. ETRI Journal 34(3), 399–409 (2012)
7. Celik, T.: Fast and Efficient Method for Fire Detection Using Image Processing. ETRI Journal 32(6), 881–890 (2010)
8. Mak, C.M., Cham, W.K.: Real-time video object segmentationin H.264 compressed domain. IET Image Process. 3(5), 272–285 (2009)

Fuzzy-Tuning PID Controller for Nonlinear Electromagnetic Levitation System

Tran Huu Luat[1], Jae-Hoon Cho[2], and Yong-Tae Kim[1,2]

[1] Dept. of Electrical, Electronic and Control Engineering, Hankyong National University
327 Joongang-ro, Anseong, Gyeonggi-do, 456-749, Korea
[2] Smart Logistics Technology Institute, Hankyong National University
167 Joongang-ro, Anseong Gyeonggi-do, 456-456-749, Korea
{tranhuuluat,jhcho,ytkim}@hknu.ac.kr

Abstract. In the paper we derive a dynamic model of the magnetic levitation system and propose a Fuzzy-Tuning PID (FTP) controller that selects the parameters of the PID controller by using fuzzy inference system. Conventional PID controller can be applied to control the electromagnet levitation. However, it is uncertain in case of load and airgap change. To solve the problem, we designed fuzzy rules of FTP considering the control response of system. We estimate the optimal parameters of PID controller through four performance indices and show the performance of PID control system in case of load and airgap response change. The performance of PID controller is compared with the proposed FTP controller. The performance of proposed system was not only faster rising time, settle time and reduced overshoot but also greater flexibility than conventional PID controller.

Keywords: Magnetic levitation system, fuzzy-tuning, PID control, fuzzy-tuning PID controller, performance indices.

1 Introduction

The magnetic levitation system without physical contact with rail has an advantage over conventional conveyors such as frictionless motion, high speed and low noise in operation. However, the stable operation of the system is a difficult problem [1-3]. The magnetic levitation system is known as high order nonlinear and open-loop unstable system [4]. The magnetic field strength of electromagnet is determined by turns of coil and the greater strength of the field depends on more turns of wire in the coil. The attraction force from a fixed magnet decreases with enlarged distance and increases at closer distances [5]. Therefore the system becomes easily unstable on varying conditions.

In order to solve the problem, stable magnetic levitation system can be archived by using a feedback controller which continuously adjusts suitable control signal for electromagnet to push back to the target position. Conventional PID controller can be applied to carry out the magnetic levitation to lift an object in a target position. However, conventional PID controller is uncertain in varying and large external

S. Kim et al. (eds.), *Soft Computing in Intelligent Control*,
Advances in Intelligent Systems and Computing 272,
DOI: 10.1007/978-3-319-05570-1_3, © Springer International Publishing Switzerland 2014

disturbance. For highly nonlinear characteristics of the magnetic levitation system, the parameters of PID controller are manually tuned under ideal condition. These parameters are not mostly suitable for the full load condition as well as external disturbance. In order to overcome these disadvantages, many researches and reports of intelligent self-tuning PID have been published such as neural network, fuzzy logic, and generic algorithm.

In the paper, we derive a dynamic model of the magnetic levitation system and design a FTP controller using fuzzy parameter tuning method and PID controller. The parameters of the PID controller are calculated by fuzzy decision system. We find the optimal parameters of PID controller through performance indices and study the performance of PID control system. The performance of conventional PID controller is compared with the proposed controller on changed load and airgap condition of the system. We verify the performance of the proposed method through simulation examples.

This paper is organized as follows. In Section 2, dynamic model of the magnetic levitation system is described. Section 3 presents PID controller with selecting the optimized parameters by performance indices and Fuzzy-Tuning PID controller. Experiment results are presented in Section 4. Finally, in section 5, draws conclusions.

2 Dynamic Model of SISO Magnetic Levitation System

Fig. 1 shows the SISO model and electrical model of the magnetic levitation system. Magnetic levitation system lifts an object by using electro magnets. The magnetic force is provided by the current through the coil which makes counterbalance with the gravitational force. The magnetic levitation system is known as two main parts: Electrical and mechanical systems.

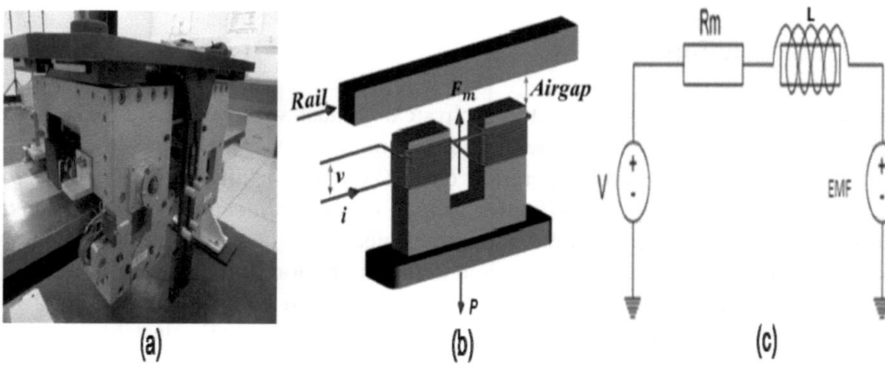

Fig. 1. Magnetic levitation system; a) Developed system, b) SISO model, b) Electrical model

Reluctance is proportional to the length and cross-sectional area of the system as

$$R_c = \frac{l_c}{\mu_c \mu_0 A_c}, \quad R_z = \frac{z}{\mu_0 A_g}, \; R_r = \frac{l_r}{\mu_r \mu_0 A_r} \tag{1}$$

where l_c is length of coil, z space between coil and rail, l_r length of rail, μ_r relative permeability of rail (H/m), μ_c permeability of magnetic coil (H/m), μ_0 permeability of free space (H/m), and A area of face, respectively. In case of $A = A_c = A_g = A_r$, total reluctance is described as

$$R_c = \frac{1}{\mu_0 A_c}\left(\frac{l_c}{\mu_c} + 2z + \frac{l_r}{\mu_r}\right) \cdot \tag{2}$$

The instantaneous voltage across a wire coil depends on the number of wire turn around core and the instantaneous variation in magnetic flux with the coil. Induced voltage e_{mf}, the magnetic linkage λ and the magnetic flux though a loop, Θ can be expressed by the Eq. (3) as

$$e_{mf} = \frac{d\lambda(t)}{dt}, \; \lambda = N\Theta, \; \Theta = \frac{Ni(t)}{R_m(z)} \tag{3}$$

where N number of coil winding. The magnetic linkage is rewritten as

$$\lambda(z,i) = N\frac{Ni(t)}{R_m(z)} = \frac{N^2 i(t)}{R_m(z)} = L_m(z)i(t) \tag{4}$$

From Eq. (2) and Eq. (4), inductance equation $L_m(z)$ is described as

$$L_m(z) = \frac{\frac{N^2\mu_0 A}{2}}{\left(\frac{l_c}{2\mu_c} + \frac{l_r}{2\mu_r} + z\right)} = \frac{\chi}{(\chi_0 + z)} \tag{5}$$

The instantaneous voltage is written as:

$$e_{mf} = \frac{\partial\lambda(i,z)}{\partial i}\frac{di(t)}{dt} + \frac{\partial\lambda(i,z)}{\partial z}\frac{dz(t)}{dt} \tag{6}$$

To rewrite the above instantaneous voltage equation

$$e_{mf} = L_m(z)\frac{di(t)}{dt} + \frac{i\partial L_m(z)}{\partial z}\frac{dz(t)}{dt} \tag{7}$$

From Eq. (7), we can find voltage equation of R-L circuit by voltage law as follows.

$$v = Ri + L_m(z)\frac{di(t)}{dt} - \frac{\chi}{(\chi_0 + z(t))^2}\frac{dz(t)}{dt} \tag{8}$$

Magnetic force produced by an electromagnet can be determined by varying of magnetic energy ψ^* as shown in Eq. (9), which moves the object respect to z axis.

$$f_m = -\frac{\partial}{\partial z'} \int \phi(t) di = -\frac{\partial \psi^*}{\partial z'}, \psi^* = \frac{1}{2} L_m(z) i(t)^2 \tag{9}$$

From Eq. (5), magnetic force is calculated as

$$f_m = \frac{\chi i(t)^2}{(\chi_0 + z')^2}, \tag{10}$$

where $z' = z_0 - z$ and z_0 is the initial position of system. Eq. (10) shows the dependence of magnetic force on the current through coil. When current is constant, attractive force will increase at closer airgap, and vice versa. In this case, the system becomes unstable at equilibrium point between gravity of force and attractive force. In order to maintain the stability of system, we have to use a feedback controller to adjust current in electromagnet varying according to airgap.

Fig. 2 shows characteristic curve of magnetic force and current of coil at different airgap intervals. The equilibrium points between the magnetic and gravity force with load 50kg are shown with the points *A (2.4A, 490N, 2mm), B (3.6A, 490N, 3mm), C (4.8A, 490N,4mm)* at airgap 4mm, 3mm and 2mm, respectively. In the paper, distance between airgap and rail is 6mm and target position of object is 3mm.

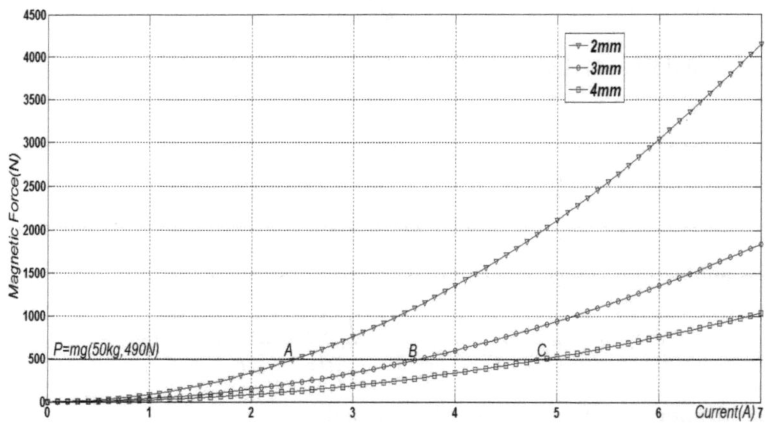

Fig. 2. Characteristic curve of magnetic force and coil current at different airgap intervals

The equation of total force can be written as

$$m\frac{d^2 z(t)}{dt} = -f_m + p_m + f_d \tag{11}$$

Total force is obtained by substituting Eq. (10) into (11) as follows.

$$m\frac{d^2z(t)}{dt}=-\frac{\chi i(t)^2}{2(\chi_0+z')^2}+mg+f_d \tag{12}$$

According to Eq. (9) and Eq. (12), the system can be described by the differential equations as:

$$\begin{cases} \dfrac{di(t)}{dt}=\dfrac{1}{L_m(z)}(v(t)+\dfrac{\chi}{(\chi_0+z)^2}\dfrac{dz(t)}{dt}-Ri(t)) \\ \dfrac{d^2z(t)}{dt}=-\dfrac{\chi i(t)^2}{2(\chi_0+z')^2}+mg+f_d \end{cases} \tag{13}$$

3 Controller Design for Magnetic Levitation System

3.1 PID Controller with Optimal Parameters

In the real magnetic levitation system, the position of electromagnet is measured by the gap sensor. In the paper, a discrete PID controller is used as shown in Eq. (14).

$$u_n(t)=u_{n-1}+K_p(e_n-e_{n-1})+\frac{K_pT_s}{2*T_i}(e_n+e_{n-1})+\frac{K_pT_d(e_n-2*e_{n-1}+e_{n-2})}{T_s} \tag{14}$$

where T_s is sampling time. Also, we select the optimal parameters of the PID controller by using the 4 performance indices: Integral of Absolute Error (IAE), Integral of Squared Error (ISE), Integral of Time Multiply Absolute Error (ITAE), and Integral of Time Multiply Squared Error (ITSE).

$$IAE=\int_0^T|e(t)|dt,\ ISE=\int_0^T e(t)^2dt,\ ITAE=\int_0^T t|e(t)|dt,\ ITSE=\int_0^T te(t)^2dt \tag{15}$$

3.2 Fuzzy-Tuning PID Controller

The magnetic levitation system easily becomes unstable in while running and is influenced by external disturbance. Therefore, it is difficult to manually tune the parameters of PID controller instantaneously. In order to solve the limitations of the PID controller, FTP controller is applied to keep the system robust under the varying condition. The parameters K'_p, K'_d and K'_i is calculated by fuzzy inference system. The varying of the parameters depends on the error between desired position and the output, and the change of error. The structure of the FTP controller is shown in the Fig. 3.

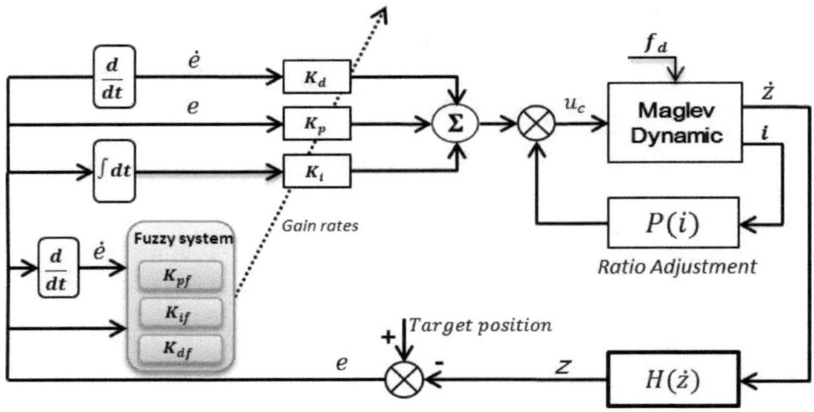

Fig. 3. Fuzzy-Tuning PID controller system

The parameters obtained from fuzzy inferences have the values in the interval [0, 1]. The gain parameters of PID controller are determined in the intervals $K_p \in [K_{p\min}, K_{p\max}]$, $K_d \in [K_{d\min}, K_{d\max}]$ and $K_i \in [K_{i\min}, K_{i\max}]$ [6-7]. Hence, K_p, K_d and K_i are evaluated by the Eq. (16) as

$$K_x = |\Delta K_x| * K_{xf} - K_{x\min}, \tag{16}$$

The membership functions of the fuzzy inputs variables are shown in Fig.4. The linguistic values of the variables are assigned as: NB: Negative Big, NM: Negative Medium, NS: Negative Small, ZO: Zero, PS: Positive Small, PM: Positive Medium, PB: Positive Big. The ranges of these input variables are [-0.003, 0.003] for e and [-0.8, 0.8] for $\dot{e}$, which are obtained from the absolute value of the system error and its change of error.

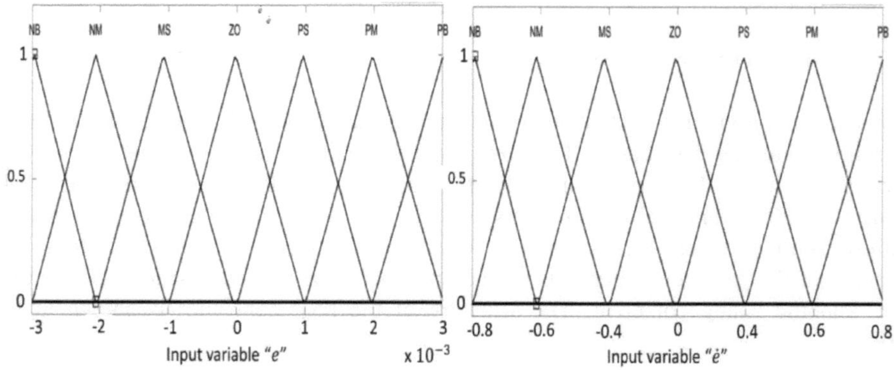

Fig. 4. Membership functions for fuzzy input variables

The membership functions of output variables, K_{pf}, K_{df} are shown in Fig. 5. The linguistic values of K_{pf}, K_{df} are assigned as: S: Small, B: Big and the linguistic value of K_{if} is assigned as: Z: Zero, S: Small, M: Medium, B: Big

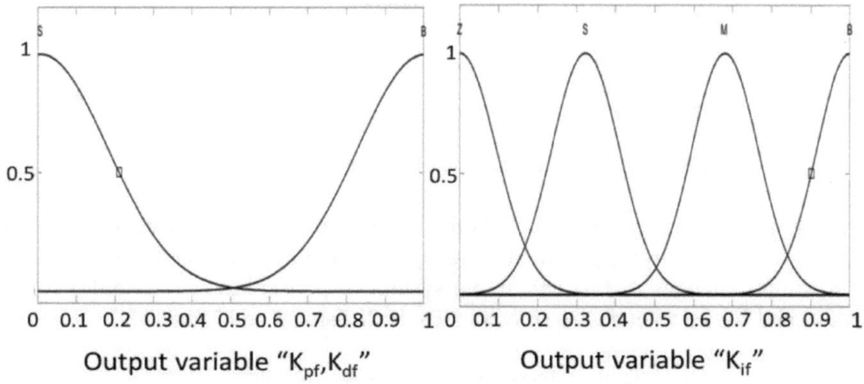

Fig. 5. Membership functions for fuzzy output variables

We use the center average defuzzifier for evaluating output results of K_{pf}, K_{df} and K_{if} as:

$$K_{xf} = \frac{\sum_{l=1}^{N} \bar{x}_l * \mu(\bar{x}_l)}{\sum_{l=1}^{N} \mu(\bar{x}_l)}, \tag{17}$$

Where N is number of fuzzy rules $\bar{x}_l$ is the center of the *l'th* fuzzy set and its height is $\mu(\bar{x}_l)$.

The control response of the magnetic levitation system is shown in Fig. 6. The curve *C* in the Fig. 6 is the simulation result of system with conventional PID controller. The typical zones on the figure depend on the polarity of the error between the target and system output with the change of error. Table 1 shows the regional classification results of the control output response. We design fuzzy rules according to the characteristics of each region. [8]

Table 1. Regional classification of control response

Number	Region	e and $\dot{e}$
1	A_1,C_2,E_1	$e>0$ and $\dot{e}<0$
2	B_1,D_2	$e<0$ and $\dot{e}<0$
3	C_1,A_2,E_2	$e<0$ and $\dot{e}>0$
4	D_1,B_2	$e>0$ and $\dot{e}>0$
5* curve (C)	A_2,B_2,C_2,D_2	$e<0$ and $\dot{e}>0$

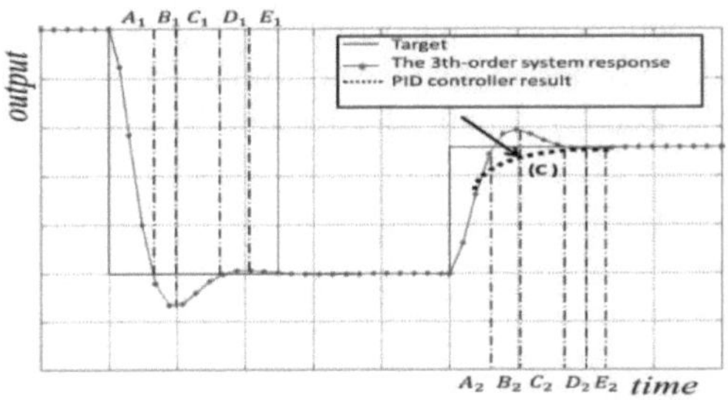

Fig. 6. The typical response of magnetic levitation system

Adjusting the parameters in such as that K_{pf} considerably increases at large airgap and vice versa, K_{if} strongly evaluates steady-state and reduce overshoot considerably and K_{df} small changes to fast and stable system response. Fuzzy rule-tables for each K_{pf}, K_{df} and K_{if} parameters are shown in Table 2, Table 3 and Table 4.

Table 2. Fuzzy rule-table for K_{pf} parameter

$e/\dot{e}$	NB	NM	NS	ZO	PS	PM	PB
NB	B	B	B	B	B	B	B
NM	S	B	B	B	B	B	S
NS	S	S	B	B	B	S	S
Z0	S	S	S	S	S	S	S
PS	S	S	B	B	B	S	S
PM	S	B	B	B	B	B	S
PB	B	B	B	B	B	B	B

Table 3. Fuzzy rule-table for K_{if} parameter

$e/\dot{e}$	NB	NM	NS	ZO	PS	PM	PB
NB	S	S	S	S	S	S	S
NM	M	M	S	S	S	M	M
NS	M	S	S	Z	S	S	M
Z0	B	M	Z	Z	Z	M	B
PS	M	S	S	Z	S	S	M
PM	M	M	S	S	S	M	M
PB	S	S	S	S	S	S	S

Table 4. Fuzzy rule-table for K_{df} parameter

$e/\dot{e}$	NB	NM	NS	ZO	PS	PM	PB
NB	B	B	B	B	B	B	B
NM	B	B	S	S	S	B	B
NS	B	S	S	S	S	S	B
Z0	S	S	S	S	S	S	S
PS	B	S	S	S	S	S	B
PM	B	B	S	S	S	B	B
PB	B	B	B	B	B	B	B

4 Experiment Results

In the paper, we apply the proposed controller to the magnetic levitation system with various load and variable target position. We use the parameters for setup of the magnetic levitation system as shown in Table 5. The parameters of PID controller are selected through the performance indices. The interval of parameters of FTP controller is estimated from working regions of the system. The controlled current through coil windings fluctuates in two cases is around 3.6A of 50kg load and 5.1A of 100kg load. Fig. 7 shows the airgap and current response of the PID control system on normal load and airgap conditions.

Table 5. Parameters of the magnetic levitation system

Parameter	Value	Dimension
Coil resistance R	4.15	Ω
Pole face area	$2200\text{x}10^{-6}$	m^2
Number of coil winding N	700	Turn
Current coefficient	266	N/A
Normal airgap	0.006	m
Normal inductance	0.22	H
Chopper DC input voltage	100	V
Mass	50,100	kg

Table 6. Performance of PID controller according to the indices

Method	Parameters	Value	M_p(%)	T_s(s)
ISE-PID	K_p=300,K_i=190, K_d=18	4.54E-05	8.186	6.043
IAE-PID	K_p=200,K_i=150, K_d=32	0.018	5.404	4.645
ITAE-PID	K_p=210,K_i=125, K_d=35	0.019	2.766	3.783
ITSE-PID	K_p=250,K_i=150, K_d=16	2.91E-05	4.911	5.020

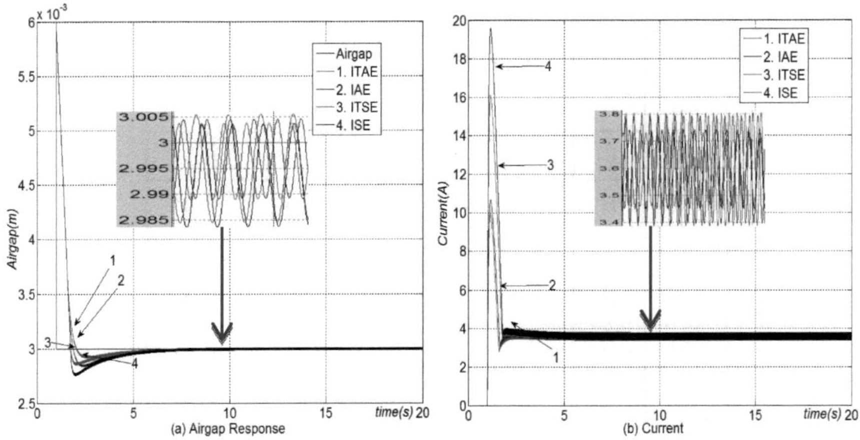

Fig. 7. The response system with PID controller on normal conditions

To give a quantitative result, Table 6 provides more the performances measuring ISE, IAE, ITAE and ITSE values. Through selecting the parameters of PID controller from the different indices is to exhibit that the magnetic levitation system still gains a fairly good response. In the four indices, ITAE-PID gives the best response with reducing overshoot and fast setting-time compare with others.

We add more 50kg loads on the system and enlarge airgap. Fig. 8 shows the airgap and current response of the PID control system with changed load and airgap. The attractive force decrease gradually when airgap is enlarged. Therefore, PID controller with before selected parameters on normal condition shows the unsteady response.

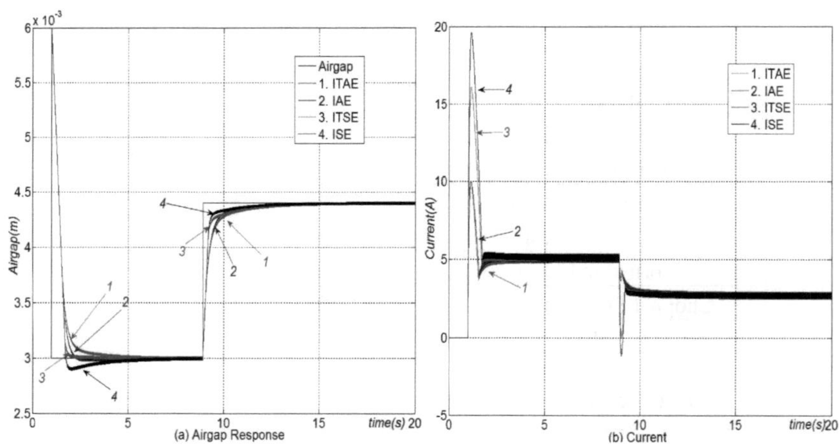

Fig. 8. The response system with PID controller on varying conditions

Fig. 9 shows the airgap and current response of FTP controller control system with normal load and airgap. Table 7 shows the performance of FTP controller according to the performance indices. In case of the normal or changed system, the proposed

FTP controller gives good performance with smaller overshoot and faster rising time and settling time than the PID controller and also maintains robustness as shown in Fig. 10.

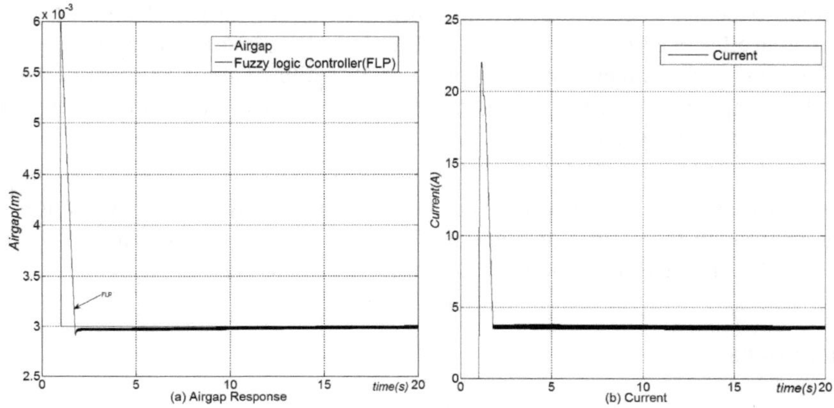

Fig. 9. The response system with FTP controller on normal conditions

Table 7. Performance of FTP controller according to the indices

Method	ISE	IAE	ITAE	ITSE	M_p(%)	T_s(s)
FTP	5.22E-05	0.031	0.129	3.92E-05	2.93	1.12

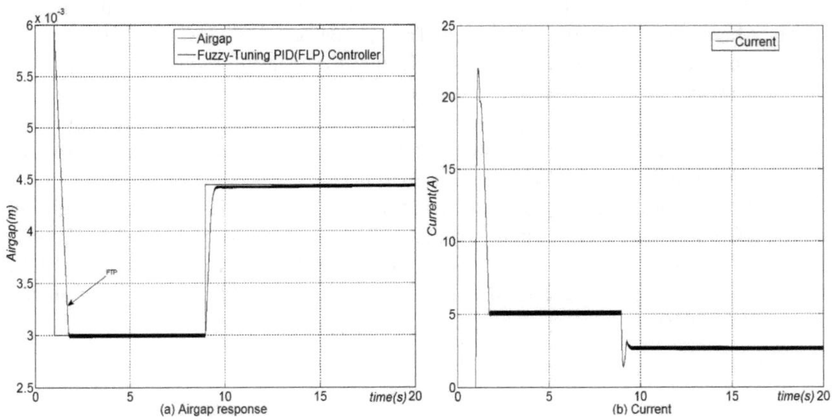

Fig. 10. The response system with FTP controller on varying conditions

Obviously, FTP controller also gives the expecting response as PID controller on in case of our system doesn't have any changes. To view Table 7 and come out an evaluation that setting-time is faster and still gain a small overshoot. Above all the others, its steady quality on changed condition of the system is depicted in Fig. 10.

5 Conclusion

In the paper, the dynamic model of the magnetic levitation system is derived and a FTP controller using fuzzy parameter tuning method is proposed. Conventional PID controller becomes uncertain in case of changed load and airgap. However, the proposed controller overcomes limitation of PID controller and has flexibility by using suitable tuning of parameters of PID controller in while running. We will continuously study the application of the proposed system for multiply input and multiply output magnetic levitation system.

Acknowledgements. This research has been supported by the Industrial Research Program through the Gyung-gi Regional Research Center (GRRC) funded by Smart Logistics Technology institute. (GRRC hankyong 2011-B01).

References

1. Alvarez-Sánchez, E., Alvarez-Gallegos, J., Castro-Linares, R.: Modeling and Controller Design of a Magnetic Levitation System. In: International Conference on Electrical and Electronics Engineering and XI Conference on Electrical Engineering, Mexico (2005)
2. HoonLee, J., Kim, Y.T.: Design of PID Controller for Magnetic Levitation RGV Using Genetic Algorithm Based on Clonal selection. The Journal of Korean Institute of Intelligent Systems 22, 239–245 (2012)
3. Groning, I., Zickermann, R., Henneberger, G.: Design of A State Control for A Solid-Coupled Magnetic Levitation Transport System. In: Proceedings of IEEE International Conference on Control Applications, vol. 1, pp. 658–661 (1999)
4. Xu, J., Zhou, Y.: ANonlinear Control Method for The Electromagnetic Suspension System of The Maglev Train. Journal of Modern Transportation 19(3), 176–180 (2011)
5. Concept of Electromagnet in Maglev Application Engineering Essay, `http://www.ukessay-s.com/essays/engineering`
6. Zulfatman, Rahmat, M.F.: Applicationof Self-Tuning Fuzzy PID Controlleron Industrial Hydraulic Actuator Using System Identification Approach. International Journal on Smart Sensing and Intelligent Systems 2(2) (2009)
7. Malessham, G., Rajani, A.: AutomaticTunning of Controller Using Fuzzy Logic. In: International Conference on Development and Application Systems, pp. 120–127 (2006)
8. Natsheh, E., Buragga, K.A.: Comparison between Conventional and Fuzzy Logic PID Controllers for Controlling DC Motors. IJCSI International Journal of Computer Science Issues 7(5) (2010)

An Effect of Traffic Speed on Maritime Accidents

Jin-Wan Park[1], Jung Sik Jeong[2], and Gyei-Kark Park[2]

[1] Department of Maritime Transportation System, Graduate School,
Mokpo National Maritime University, Mokpo City, 530-752, Republic of Korea
pjinwan@mmu.ac.kr
[2] Division of International Maritime Transportation Science,
Mokpo National Maritime University, Mokpo City, 530-752, Republic of Korea
{jsjeong,gkpark}@mmu.ac.kr

Abstract. VTS has been operated at the ports and coastal areas in order to prevent maritime accidents and improve transport efficiency, as described in IMO Resolution A.857(20), Guideline for VTS. To reduce the ship collision, we need to understand the factors of the traffic characteristics which is related with collisions. Although many considerable studies have been conducted over the past several years to prevent ship collisions, the results are not practical. We investigated the correlation among statistical parameters of vessel speed, traffic flow, and potential collision rate in VTS area. Finally, This research provides persuasive evidence to enforce an safe speed to reduce ship collisions in VTS area.

Keywords: VTS, Marine Accidents, Safe Ship Speed, Collision Rate, Maritime Traffic.

1 Introduction

The movement of goods at sea has supported world-wide shipping for centuries, giving rise to a need for ships to navigate safely and efficiently. To this end, maritime authorities throughout the world have provided aids to navigation(AtoN) in and around their coastal waters. As part of the AtoN, the Vessel Traffic Service(VTS) spread out to the world with starting from the port of Douglas(Isle of Man) and Liverpool in 1948[2]. The VTS that defined in the IMO Resolution A.857(20) is a service implemented by a Competent Authority, designed to improve the safety and efficiency of vessel traffic and to protect the environment[1]. The service should have the capability to interact with the traffic and to respond to traffic situations developing in the VTS area that is the delineated, formally declared service area of the VTS[1]. Although the VTS had contributed to make safety of navigation for several years, the authorities of VTS effort to develop and advance the countermeasures continuously to remove possibility of a ship collision. Because 35% of all accidents in the VTS area during the period from 2006 to 2010 are the ship collisions and it shown the highest

S. Kim et al. (eds.), *Soft Computing in Intelligent Control*,
Advances in Intelligent Systems and Computing 272,
DOI: 10.1007/978-3-319-05570-1_4, © Springer International Publishing Switzerland 2014

percentage[4]. Because only one occurrence of ship collision can make a huge casualties and property damage and marine pollution, competent authority, as well as VTS authority implement the many countermeasures that separated by ship type, weather conditions, seasons to prevent the ship collisions. The safe speed appears as main content in these many countermeasures. According to the COLREG (International Regulations for Preventing Collisions at Sea, 1972), a safe speed ensures the safety of ship is heavily influenced by environmental conditions, so that human's subjective judgment and experience to be considered majority opinion in order to derive the range of safe speed[3]. A number of studies on the safe speed have been conducted to identify the range of safe speed in many countries[4]. We also already conducted research about an effect of ship speed on the maritime accidents[5]. However, the results of them were not used in practice because of lack of the effectiveness.

In this study, we proposed the mathematical model to define the range of safe speed in particular traffic lane. The mathematical model was represented by the multiple linear regression and the multivariate ratio of polynomials. In order to obtain the best model that fits the observed data, two measurements have been applied in this research, the coefficient of determination (R^2) and Akaike's Information criterion(*AIC*). All the data of maritime traffic and accidents were obtained from VTS. To understand any relationship between potential collision rate(*Cr*) and statistical distribution of ship speed at three traffic lanes, we investigate the mean speed(*MEAN*), standard deviation of speed(*SD*), flow per route(*FPR*) at each traffic route.

2 Potential Collision Rate of Traffic Route

In the field of road traffic, many studies have conducted to improve the safety on highways and understand the factor related with crash rate[6]. It makes easy to identify and implement effective countermeasures to reduce crash occurrences. Since the number of crash is dependent on the length of highway, the hours for which the data are collected, and the number of lanes, it is necessary to normalize the number of crash as a crash rate[7]. In this study, we used the equation with respect to the potential collision rate(C_r) at a traffic route. That is defined by equation (1)[6].

$$C_r = \frac{C}{L \times W \times H} \tag{1}$$

where C_r denotes potential collision rate of the traffic route of interest, C is the number of incidents of collision, L is route length, W is width of route, and H is the time period of the data collected. The number of incidents of collision was investigated in the light of the *CPA*(Closest Point of Approach) and the *TCPA*(Time to *CPA*) between the all ships of the object traffic route. We fixed the no more than 0.15NM of *CPA* and the no more than 3min of *TCPA* for regard them as incident of collision. The *CPA* of two vessels help you if you want to understand the risk of collision associated with the International maritime collision prevention rules(COLREGS). When one vessel's current position, course and speed is given by (x, y, θ, v) and the other vessel's current position, course and speed are given by $(x_1, y_1, \theta_1, v_1)$, $TCPA(t_{\min})$ have to be obtained first to calculate the *CPA*. Generally, for any given time t (where the current time is t=0), and assuming the other vessel stays on a constant trajectory, the distance between the two vessels is as follows[7].

$$dist^2(\theta, v, t) = a_2t^2 + a_1t + a_0 \tag{2}$$

where

$$a_2 = cos^2(\theta)v^2 - 2\,cos(\theta)\,vcos(\theta_1)v_1 + cos^2(\theta_1)v_1^2 + sin^2(\theta)v^2 - 2\,sin(\theta_1)\,v_1 + sin^2(\theta_1)v_1^2$$

$$a_1 = 2\,cos(\theta)\,vy - 2\,cos(\theta)\,vy_1 - 2ycos(\theta_1)v_1 + 2\,cos(\theta_1)\,v_1y_1 + 2\,sin(\theta)\,vx - 2\,sin(\theta)\,vx_1 - 2xsin(\theta_1)v_1 + 2sin\,(\theta_1)v_1x_1$$

$$a_0 = y^2 - 2yy_1 + y_1^2 + x^2 - 2xx_1 + x_1^2.$$

The $TCPA(t_{\min})$ is obtained by differentiating the equation (2), and determining the distance value become 0. Therefore the equation of the *TCPA* is as follows[7].

$$t_{\min} = \frac{-a_1}{2a_2}$$

The actual *CPA* value is then obtained by substituting the *TCPA* into the equation (2). The equation of the *CPA* is as follows[7].

$$CPA(\theta, v, t) = \sqrt{a_2t_{min}^2 + a_1t_{\min} + a_0} \tag{3}$$

Based on the log data of VTS collected, the values of *CPA* are obtained by substituting the value of vessel's position, course and speed into the equation (3).

3 Data Collection

We obtained VTS's log data for speed and traffic flow from 3 September to 5 September 2012. It is obtained by VLR(VTS Logging and Replay) system which stores the RADAR images, the tracking information and the voice of VHF equipment in the MOKPO VTS center. The next step of this task is to collect the ship collisions data in VTS area from 2006 to 2011, also geometric data such as length and width of each traffic route. To investigate any relationship between ship speed and maritime accidents, the traffic routes where collision accidents occurred over the last 6 years were surveyed as in Fig. 1.

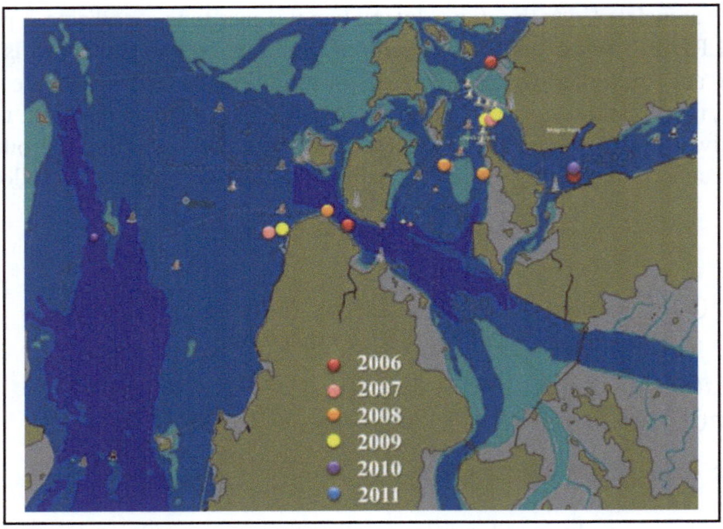

Fig. 1. Area of ship's collisions

Base on the data of collision accidents, we selected the object traffic routes with considering the frequency of collision accidents. Fig. 2 shows the three traffic routes. The collision accidents occurred most frequently in these traffic routes.

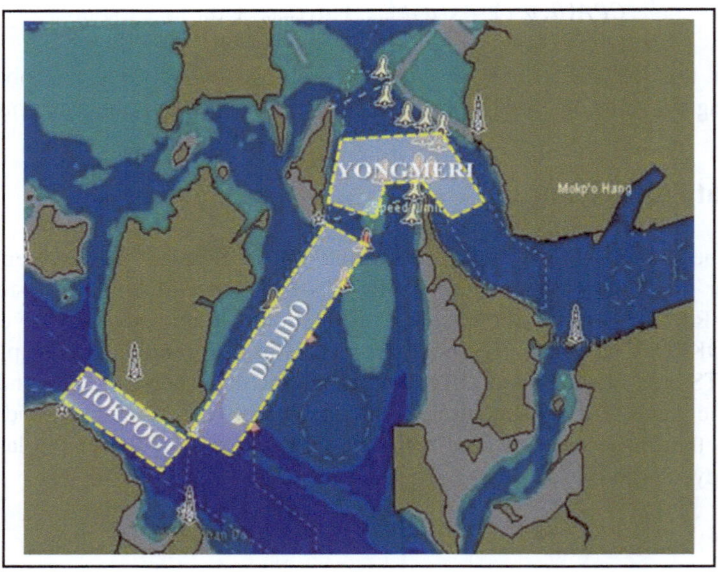

Fig. 2. Traffic routes of analysis object

Table 1 shows length, width and the number of collision accidents for the three traffic routes. The YONGMERI that is not only a curved and marine bridge is also installed has the most of the number of collision accidents.

Table 1. Data collected by traffic route

TRAFFIC ROUTE	LENGTH	WIDTH	COLLISION NO.
YONGMERI	1.4NM	0.22NM	3
DALIDO	1.6NM	0.3NM	1
MOKPOGU	1.05NM	0.24NM	2

In order to facilitate an analysis of the data, we arrange the ship's log data by traffic route. Table 2 shows the part of the data. The log data arranged be used in obtain the value of each variables.

Table 2. Log data of VTS

Vessel	Latitude	Longitude	Course (°)	SPD (knot)	Date	Time
A	N 34-46.1346	E 126-17.883	106.1	10.7	2012-9-3	15:57:20
B	N 34-45.5564	E 126-19.3211	42.3	13.2	2012-9-4	8:41:0
C	N 34-47.3595	E 126-17.8836	277.9	4.8	2012-9-5	10:18:20
D	N 34-48.4331	E 126-13.2357	164	9	2012-9-5	23:43:20

Once all the data are collected and filtered, they are entered into a spreadsheet for analysis. The spreadsheet consisted of following variables: mean speed(knot), flow per route in vessels per hour(vph), standard deviation of speed(knot), potential collision rate. The potential collision rate is calculated by dividing the number of crashes by length, number of hours the data, and route width(m). All values of the variables are calculated hourly. Table 3 shows the variables data for traffic route of DALIDO.

Table 3. Traffic route data of DALIDO

HOURS	*MEAN* (kont)	*FPR* (vph)	*SD* (knot)	*Cr* (No. of incidents/hr/NM ^{2})
1	9.219375	6	1.91786	4.166667
2	7.705634	5	1.878821	4.166667
3	9.223034	7	0.856291	NaN
4	7.384862	5	1.234184	4.166667

4 Model Developement

In order to develop the correlation model between the potential collision rate and the ship speed, we analyzed data using with the Multiple Linear Regression and the Multivariate Ratio of Polynomials. All modeling was done using the Number Cruncher Statistical Systems(NCSS) software.

4.1 Multiple Linear Regression

The data were first applied to the multiple linear regression, which expresses the mean of the dependent variable as a linear correlation of two or more independent variables. The models were fitted to the data by the method of least squares and t-test performed to test the resulting regression coefficients against the null hypothesis of being equal to zero at the 5 percent significance level. The adequacy of the fit was judged based on the R^2 value and the normality of the residuals[10]-[13].

4.2 Multivariate Ratio of Polynomials

The multivariate ratio of polynomial procedure searches through hundreds of potential curves looking for the model that fits data the best. The models developed through this process offer a larger variety of surfaces than the usual polynomial models. In addition, the model should be studied graphically to determine that the model behaves as expected between data points. The program, NCSS, contains a function called the multivariate ratio of polynomials search. This function provides a shortcut to the normally slow iterative process whereby an approximate solution is found quickly because a large number of models may be searched in a short period of time. The procedure provides the option to enter one dependent and up to four independent variables. Both the dependent and independent variables may be transformed in following forms for searching: $1/Y^2$, $1/Y$, $1/\sqrt{Y}$, ln(Y), Y^2. The procedure also provides options for which models may be searched. The models searched in this research involved only the numerator polynomial of the hierarchical type of model up to the second power. Variables are included in a model, the number of free parameters, the variable transformation and the normality of the residuals are reviewed in each models searched by NCSS, and utilizing the *AIC* value and the R^2 value as the criterion for selection of the best model. NCSS uses a pseudo R^2 value instead of the R^2 value as the determination coefficient for non-linear regression. Pseudo R^2 formula is as follows[6][13].

$$R_p^2 = \frac{Model\ SS - Mean\ SS}{Total\ SS - Mean\ SS} \tag{4}$$

where *Mean SS*=$\sum(\bar{y})^2$, $\bar{y}$= mean of the observations, *Model SS*=$\sum(\hat{y}_i)^2$, $\hat{y}_i$= model estimates, *Total SS*=$\sum(y_i)^2$, y_i=actual observation.

In addition, *AIC* was used as the criterion for selection of best non-linear model. AIC formula is as follows[8].

$$AIC = -n\ln\left(\frac{SSE}{n-k}\right) + 2k \quad (5)$$

where $SSE=\sum(y_i - \hat{y}_i)^2$, k = number of free parameters in the model, and n = number of model residuals. The first term in the *AIC* equation measures the badness of fit, or bias, when the maximum likelihood estimates of the parameters are used. The second term measures the complexity of the model, thus penalizing the model for using more parameters. The goal for selecting the best model is therefore a minimization of the criterion, thus selecting the best fit with the least complexity[6].

5 Results

5.1 Results of the Multiple Linear Regression Analysis

The multiple linear regression analysis was conducted to define the relationship between the potential collision rate(C_r) and independent variables including the mean speed(MEAN), standard deviation of speed(SD) and flow per route(FPR) in each traffic routes.

1) Traffic Route of the YONGMERI

Firstly, we assessed the multicollinearity in the data of the YONGMERI traffic route. The multicollinearity was not defined by measuring the variance inflation factor (VIF). Therefore the multiple linear regression analysis was conducted with all independent variables *MEAN*, *SD*, and *FPR* to predict the C_r. The results of analysis are that the *FPR*(p=0.000<0.05) have a impact on the C_r and the power of explanation of the best model is 69.5% as shown in the Table 4. The estimated model is:

$$C_r=8.385+2.403*FPR-2.918*MEAN+1.090*SD$$

Table 4. Regression Coefficients for YONGMERI

Independent Variable	B	SE	β	T	p
Intercept	8.385	14.231	-	0.589	0.558
FPR	2.403	0.232	0.846	10.341	0.000
MEAN	-2.918	1.555	-0.142	-1.876	0.066
SD	1.090	1.908	0.045	0.571	0.570
	R^2=0.711	R^2 (adj)=0.695	F=44.352	p=0.000	

The Table 4 reports the values and significance tests of the regression coefficients. The **B** is the values of regression coefficients. They are the least squares estimates of the parameters. The **SE** is the value of the standard error of regression coefficient. And the ß is the value of the standardized coefficient and the **T** is the t-test value for testing the hypothesis that **B**=0 versus the alternative that **B**≠0 after removing the influence of all other independent variables[13]. The **p** is the p-value for the significance test of the regression coefficient. The results of test of the normality of the residuals by the Shapiro-Wilk verification method are that the statistics of the residuals standardized is 0.973 and significant probability is 0.222, it is greater than 0.05, it was meet the normality of the residuals.

2) Traffic Route of the DALIDO

The multicollinearity was also not defined. The results of analysis are shown as Table 5. The *FPR*(p=0.000<0.05) and *MEAN*(p=0.025<0.05) have a impact on the C_r and the power of explanation of the best model is 55.5%. The estimated model is:

$$C_r=3.585+2.168*FPR-1.261*MEAN-0.211*SD$$

Table 5. Regression Coefficients for DALIDO

Independent Variable	B	SE	ß	T	P
Intercept	3.585	5.319	-	0.674	0.503
FPR	2.168	0.263	0.772	8.260	0.000
MEAN	-1.261	0.546	-0.222	-2.310	0.025
SD	-0.211	0.694	-0.030	-0.304	0.762
	R^2=0.579	R^2 (adj)=0.555	F=23.841	p=0.000	

The results of test of the normality of the residuals by the Shapiro-Wilk verification method are that the statistics of the residuals standardized is 0.936 and significant probability is 0.005, it is less than 0.05, it was not meet the normality of the residuals.

3) Traffic Route of the MOKPOGU

The multicollinearity was also not defined by measuring the variance inflation factor (VIF). The results of analysis are that the *FPR*(p=0.000<0.05) and *MEAN*(p=0.000<0.05) have a impact on the C_r and the power of explanation of the best model is 55.5% as shown in the Table 6. The estimated model is:

$$C_r=14.856+2.442*FPR-2.186*MEAN+0.853*SD$$

Table 6. Regression Coefficients for MOKPOGU

Independent Variable	B	SE	ß	T	P
Intercept	14.856	5.852	-	2.539	0.014
FPR	2.442	0.361	0.669	6.765	0.000
MEAN	-2.186	0.503	-0.413	-4.345	0.000
SD	0.853	0.780	0.108	1.093	0.280
	R^2=0.579	R^2 (adj)=0.552	F=21.987	p=0.000	

The results of test of the normality of the residuals by the Shapiro-Wilk verification method are that the statistics of the residuals standardized is 0.975 and significant probability is 0.352, it is greater than 0.05, it was meet the normality of the residuals.

5.2 Results of the Multivariate Ratio of Polynomials

1) Traffic Route of the YONGMERI

The best model is defined by lowest value of *AIC* and highest value of R_p^2. The estimated model is as follows.

$$C_r=((270.6627)-(311.4897)*(ln(MEAN))+(86.32126)*(ln(MEAN))^2+(192.28)*(1/(SD*SD))-(85.36794)*(ln(MEAN))*(1/(SD*SD))-(0.04254984)*(1/(SD*SD))^2+(0.4870574)*(FPR*FPR)-(0.1828188)*(ln(MEAN))*(FPR*FPR)-(0.1121868)*(1/(SD*SD))*(FPR*FPR)-(1.92984E\text{-}05)*(FPR*FPR)^2).$$

The value of R_p^2 for the model is 0.808, and *AIC* is -280.293. Examination of the normal distribution of residuals was conducted using the normal probability plot of residuals. The residuals are normally distributed. The details of results are as Table 7.

Table 7. Results for YONGMERI

SSE	K	N	R_p^2	AIC
8507.133	10	58	0.808	-280.293

The graphical representations of the model are shown in Fig. 3, 4, and 5. Fig. 3 shows the correlation between the potential collision rate and the standard deviation of speed when the value of mean speed is 8.56knot and the value of flow per route is 17vph. The potential collision rate increases as the standard deviation of speed increases. The rate of increase in potential collision rate is very high until the standard deviation of speed of about 4knot is reached, after that the rate of increase get lower.

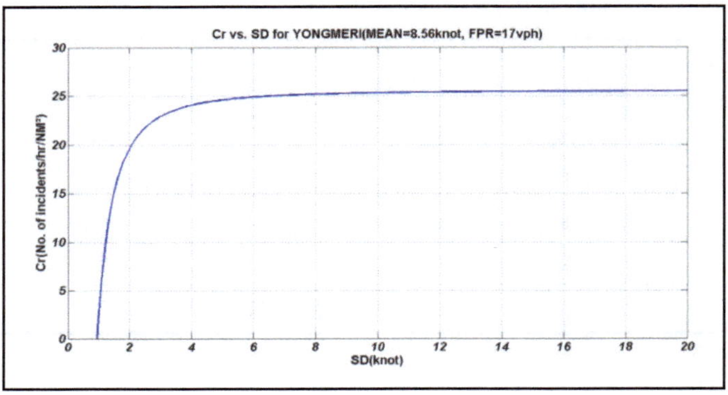

Fig. 3. Cr vs. *SD* for YONGMERI

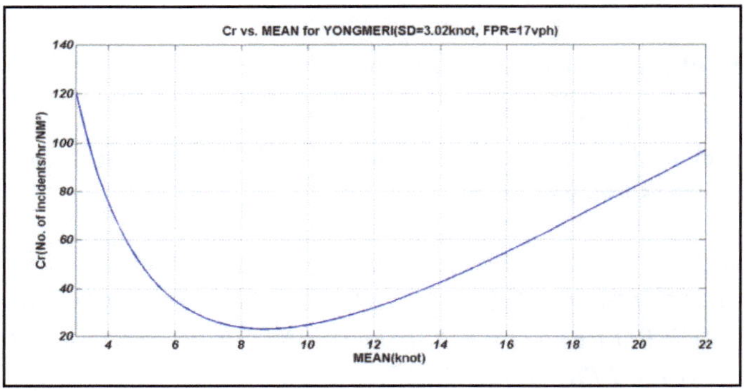

Fig. 4. Cr vs. *MEAN* for YONGMERI

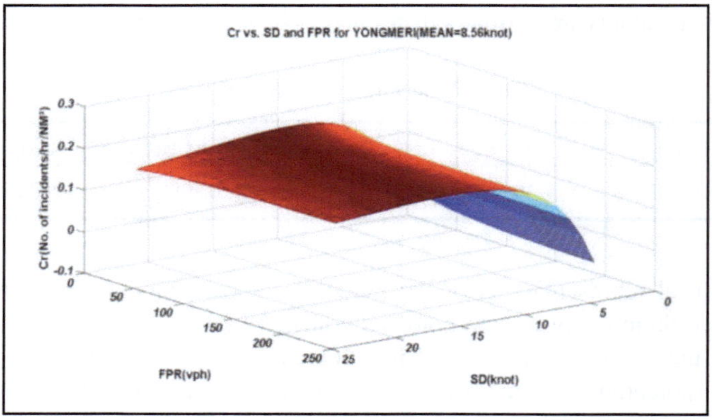

Fig. 5. Cr vs. *SD* and *FPR* for YONGMERI

Fig. 4 shows the correlation between the potential collision rate and the mean speed when the value of standard deviation of speed is 3.02knot and the value of flow per route is 17vph. The potential collision rate is lowest when the value of mean speed is between about 8knot and 10knot. Fig. 5 shows the correlation between the potential collision rate and the standard deviation of speed, and flow per route when the mean speed is 8.56knot. The potential collision rate increases as the standard deviation of speed increases until the standard deviation of speed reach about 7knot, and then the potential collision rate decreases gradually. The flow per route increases as the potential collision rate increases.

2) Traffic Route of the DALIDO

The best model is defined by lowest value of *AIC* and highest value of R_p^2. The estimated model is as follows.

$$C_r = ((-16.61025)+(5.112154)*MEAN-(0.4274818)*MEAN^2-(4.578855)*\sqrt{SD}+(2.129504)*MEAN*\sqrt{SD}-(5.501996)*\sqrt{SD}^2+(0.2298801)*(FPR*FPR)-(0.01572835)*MEAN*(FPR*FPR)+(0.02220303)*\sqrt{SD}*(FPR*FPR)-(9.798831E-05)*(FPR*FPR)^2).$$

The value of R_p^2 for the model is 0.681, and *AIC* is -211.175. Examination of the normal distribution of residuals was conducted using the normal probability plot of residuals. The residuals are normally distributed. The details of results are as Table 8.

Table 8. Results for DALIDO

SSE	K	N	R_p^2	AIC
2854.834	10	56	0.681	-211.175

The graphical representations of the model are shown in Fig. 6, 7, and 8. Fig. 6 shows the correlation between the potential collision rate and the standard deviation of speed when the value of mean speed is 9.33knot and the value of flow per route is 8vph. The potential collision rate increases as the standard deviation of speed increases up to about 2knot, but then, the potential collision rate was rapidly decreased as the standard deviation increases. Fig. 7 shows the correlation between the potential collision rate and the mean speed when the value of standard deviation of speed is 3.54knot and the value of flow per route is 8vph. It shaped the curve of parabolic overall, potential collision rate had the maximum value when the value of mean speed was about 9knot. Fig. 8 shows the correlation between the potential collision rate and the standard deviation of speed, and flow per route when the mean speed is 9.33knot. The potential collision rate increases as the standard deviation of speed increases until the standard deviation of speed reach about 10knot, and then the potential collision rate decreases gradually. The flow per route increases as the potential collision rate increases.

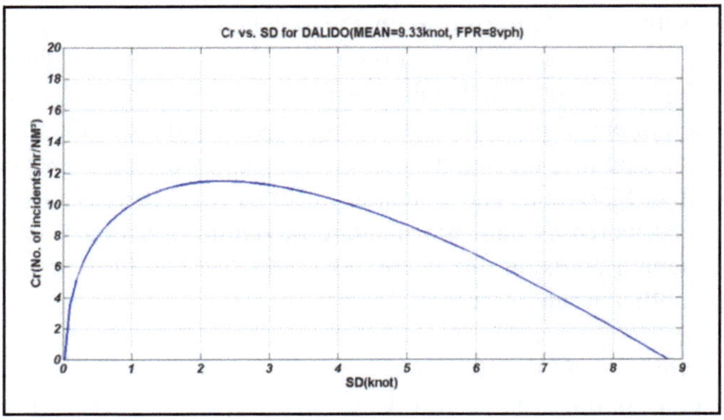

Fig. 6. Cr vs. *SD* for DAFLDO

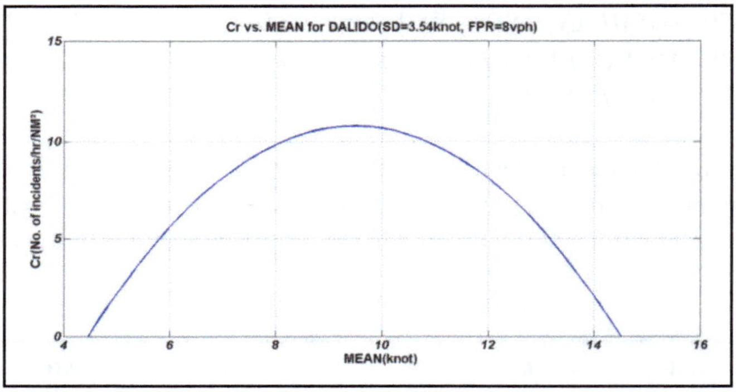

Fig. 7. Cr vs. *MEAN* for DALIDO

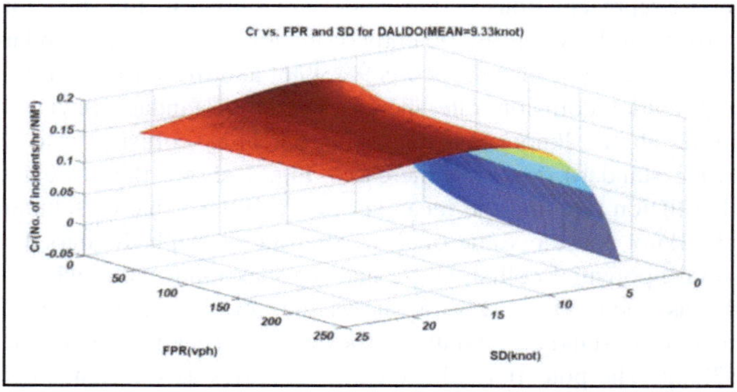

Fig. 8. Cr vs. *FPR* and *SD* for DALIDO

3) Traffic Route of the MOKPOGU

The best model is defined by lowest value of *AIC* and highest value of R_p^2. The estimated model is as follows.

$$C_r = ((10.36267)-(1730.331)*(1/(MEAN*MEAN))-(2624.022)*(1/(MEAN*MEAN))^2+(22.07549)*(1/(SD*SD))+(1529.092)*(1/(MEAN*MEAN))*(1/(SD*SD))-(4.408104)*(1/(SD*SD))^2-(7.232133)*\sqrt{FPR}+(748.8388)*(1/(MEAN*MEAN))*\sqrt{FPR}-(15.56421)*(1/(SD*SD))*\sqrt{FPR}+(2.676923)*\sqrt{FPR}^2$$

The value of R_p^2 for the model is 0.648, and *AIC* is -215.559. Examination of the normal distribution of residuals was conducted using the normal probability plot of residuals. The residuals are normally distributed. The details of results are as Table 9.

Table 9. Results for MOKPOGU

SSE	K	N	R_p^2	AIC
3895.788	10	52	0.648	-215.559

The graphical representations of the model are shown in Fig. 9, 10, and 11. Fig. 9 shows the correlation between the potential collision rate and the standard deviation of speed when the value of mean speed is 8.85knot and the value of flow per route is 7vph. The potential collision rate increases as standard deviation of speed increases up to about 4knot, and then, the potential collision rate have the stationary value.

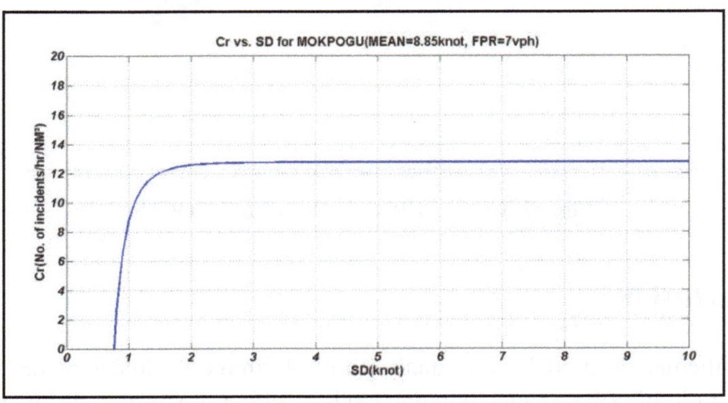

Fig. 9. Cr vs. *SD* for MOKPOGU

Fig. 10 shows the correlation between the potential collision rate and the mean speed when the value of standard deviation of speed is 3.98knot and the value of flow per route is 7vph. The potential collision rate increases as the mean speed increases until the mean speed of about 4knot is reached, after that the potential collision rate begins to decrease. Fig. 11 shows the correlation between the potential collision rate and the standard deviation of speed, and flow per route when the mean speed is 8.85knot. The potential collision rate increases as the standard deviation of speed increases until the standard deviation of speed reach about 9knot, and then the potential collision rate decreases gradually. The flow per route increases as the potential collision rate increases.

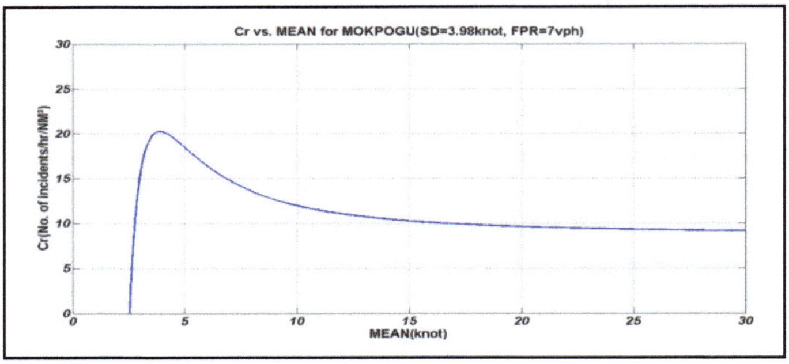

Fig. 10. Cr vs. *MEAN* for MOKPOGU

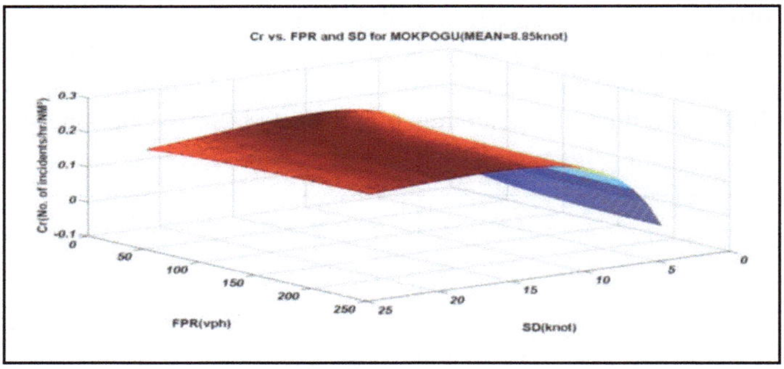

Fig. 11. Cr vs. *FPR* and *SD* for MOKPOGU

6 Conclusions

The two mathematical models were analyzed by the multiple linear regression and the multivariate ratio of polynomials to understand the effect of speed and flow on potential collision rate for different types of traffic route in VTS area. The Multiple linear regression models showed the significance between the potential collision rate and the speed, and traffic flow. The flow per route was shown the highest significance among the three variables in all traffic routes. And the mean speed has shown the significance in the DALIDO and MOKPOGU traffic route. The models that were developed by the multivariate ratio of polynomials portrayed the details of effects of speed on the potential collision rate. However, the results show that the potential collision rate is not dependent on any single independent variable, but on the complex interaction of all the independent variables.

In this study, the proposed regression model shows the effect of mean speed, standard deviation of speed and flow per route on the potential collision rate. However, these correlations are very complex. In order to derive a more accurate result, it is necessary to collect much more data. In particular, the size and type of the ship should be considered to derive the collision rate of the vessel. Further, in the sea, it is necessary to derive the safe speed by distinguishing the restricted visibility and insight of one another. The proposed safe speed is open. The proposed model needs to be studied in terms of ship's dynamics.

Acknowledgment. This research is a part of the project titled "Development of the Intelligent Technology for Marine Accident Protection & Salvage" funded by the Ministry of Oceans and Fisheries, Republic of Korea.

References

1. IMO, Guidelines for Vessel Traffic Services. Res. A 857(20), 3–7 (1997)
2. IALA, Vessel Traffic Services Manual, 5th edn., pp. 16–18 (2012)
3. IMO, International Regulations for Preventing Collisions at Sea 1972 (2003)
4. Ministry of Oceans and Fisheries, Statistics of Maritime Accidents for Vessel Traffic Service, Navigational Safety Information-896 (April 2011)
5. Park, J.W., Jeong, J.S., Park, G.K.: A Study on Influence of Ship Speed and Flow in Maritime Accidents. In: ACMSSR 2013, pp. 112–115 (August 2013)
6. Garber, N.J., Ehrhart, A.A.: The Effect of Speed Flow and Geometric Characteristics on Crash Rates for Different Types of Virginia Highways. Virginia Transportation Research Council, FHWA/VTRC 00-R15, 1–7 (2000)
7. Benjamin, M.R., Leonard, J.J., Curcio, J.A., Newman, P.M.: A Method for Protocol-Based Collision Avoidance Between Autonomous Marine Surface Craft. Journal of Field Robotics 23(5), 338–339 (2006)
8. Akaike, H.: A New Look at the Statistical Model Identification. IEEE Trans. on Automatic Control 19(6), 716–723 (1974)
9. Lee, E., Heo, Y.B., Hong, S.B., Kim, J.O., Jeong, K.N., Park, S.T., An, G., Yun, J.S.: Theory of Vessel Traffic System, pp. 20–21. HAEIN Pubilsher (2005)
10. Park, B.J.: Modern statistical theory and application using EXCEL, Seoul. SIGMAPRESS Ltd., 311–313 (2004)
11. Ministry of Oceans and Fisheries, Local Regulation for Vessel Traffic Service, Regional Maritime Affairs and Port Office, MOF, No. 106 (August 2013)
12. Kim, Y.D., et al.: Statistics using SPSS, pp. 294–295. Seoul National University, JAYU Academy, Paju (2009)
13. Hintze, J.L.: NCSS User's GuideⅢ, ch. 305, 371, 376. NCSS, Kaysville (2007)

Robust Object Detection in Sea Environment Based on DWT

Jongmyeon Jeong[1], Ki Tae Park[2], and Gyei-Kark Park[1,*]

[1] Department of Computer Engineering, Mokpo National Maritime University, South Korea
[2] Center for Integrated General Education, Hanyang University, South Korea

Abstract. In this paper, a new method for detecting various objects that can be risks to safety navigation in sea environment is proposed. By analyzing Infrared(IR) images obtained from various sea environments, we could find out that object regions include both horizontal and vertical direction edges while background regions of sea surface mainly include vertical direction edges. Therefore, we present an approach to detecting object regions considering horizontal and vertical edges. To this end, in the first step, image enhancement is performed by suppressing noises such as sea glint and complex clutters using a statistical filter. In the second step, a horizontal edge map and a vertical edge map are generated by Discrete Wavelet Transform. Then, a saliency map integrating the horizontal and the vertical edge maps is generated. Finally, object regions are detected by an adaptive thresholding method.

Keywords: safety navigation, horizontal edge map, vertical edge map, discrete wavelet transform, object region detection.

1 Introduction

Over the past few decades, ships have been developed for achieving economies of scale which become larger, more speed, and safer. However, because the large size and high-speed of ships can cause extensive damage in marine accidents, it is necessary to consider analysis of risk factors and measures for the prevention of accidents [1]. According to [1-3], the accident type with the highest probability in sea environment is a collision between ships and the main cause of the accident is the lack of observing surrounding environments. In ship operation, the observation of surrounding environments is the most important factor. A ship is operated by sea route and steering of a captain. Therefore, most of collision occurs by artificial decisions, except uncontrollable situations. The observation in ship operation is categorized into two types: (1) eye observation; (2) observation based on various equipments such as synthetic aperture radar (SAR) and infrared (IR) camera. When operating a ship, ship operators must consider both the two types because eye observation is limited under various surrounding environments such as bad weather, haze, and night navigation. SAR is widely utilized for obtaining information of surrounding environment.

* Corresponding author.

S. Kim et al. (eds.), *Soft Computing in Intelligent Control*,
Advances in Intelligent Systems and Computing 272,
DOI: 10.1007/978-3-319-05570-1_5,

However, operators can be confused owing to some noises and data errors based on multiple reflections [4]. On the contrary, since IR camera can obtain identical IR images under the poor conditions above mentioned, objects which are the causes of sea accident can be automatically detected. Therefore, IR camera can make up for the weak points of both eye observation and SAR [5-7]. In this paper, for safety navigation of a ship, a new object detection technique by analyzing IR images is proposed.

2 The Proposed Method

By analyzing IR images captured under sea environments, we find out that object regions include both the vertical and the horizontal edges while regions with the gradient of horizontal direction mainly appear in the sea surface and the horizontal line. In this paper, a new approach to scale invariant object detection considering direction of edges in sea environment is proposed. To this end, in the first step, noises reduction through image enhancement is performed for effectively detecting edges. In the second step, the horizontal and vertical edges are detected by the discrete wavelet transform(DWT). Consequently, two edge maps are generated. Finally object regions are detected by combining the two edge maps and threshold technique. Figure 1 shows the overall flowchart of the proposed method.

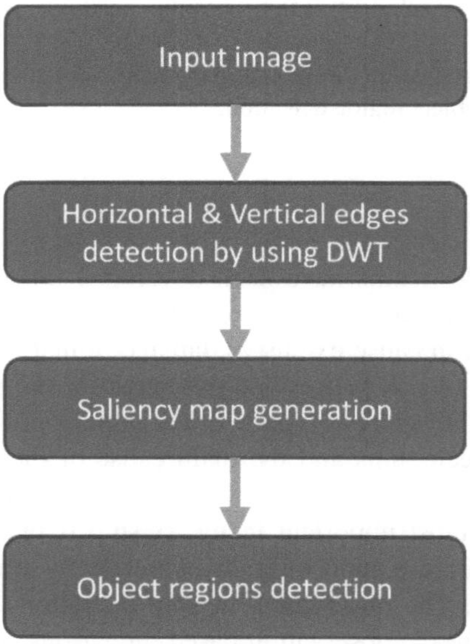

Fig. 1. Overall flowchart of the proposed method

2.1 Noise Reduction by Morphological Operation

Because IR images captured under sea environment include various noises such as sea glints and clutters, the accuracy of object detection can be degraded. Therefore, the method for image enhancement to remove various noises is required. In this paper, in order to effectively remove noises, we utilize morphological operations. Erosion that is morphological operation based on the set theory is defined as (1) [8].

$$A - B = \{x \mid (B)_x \subset A\} \tag{1}$$

where A and B are the binarized image extracting candidate object region and the structuring element, respectively. Noises are eliminated and object regions are eroded in image A by (1). On the contrary, dilation dilates the object regions and noises as shown in (2).

$$A \oplus B = \{x \mid [(\hat{B})_x \cap A] \subseteq A\} \tag{2}$$

Opening and closing operations are defined by combining erosion and dilation operations. As shown in (3), the opening of A by B is obtained by the erosion of A by B, followed by dilation of the resulting image by B. The closing of A by B is obtained by the dilation of A by B, followed by erosion of the resulting structure by B in (4).

$$A \circ B = (A \ominus B) \oplus B \tag{3}$$

$$A \bullet B = (A \oplus B) \ominus B \tag{4}$$

The erosion operation effectively eliminates noise regions by eroding the outside of the object and the dilation operation enlarges the object. The opening operation generally makes the shape of the object round and eliminates the protruding regions. Moreover, this operation is effective for removing noises. Therefore, in this paper, we perform the morphological opening operation in order to effectively remove noises. Fig. 2 shows the enhanced results by the morphological operations.

(a) An input image (b) An enhanced image

Fig. 2. Image enhancement result by morphological operations

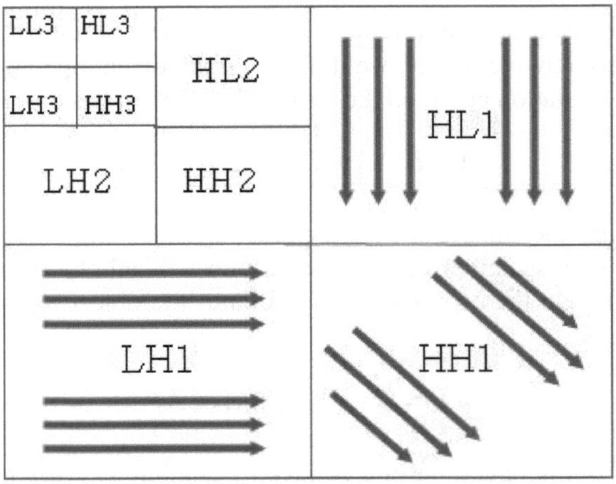

(a) The scheme of wavelet decomposition into three levels

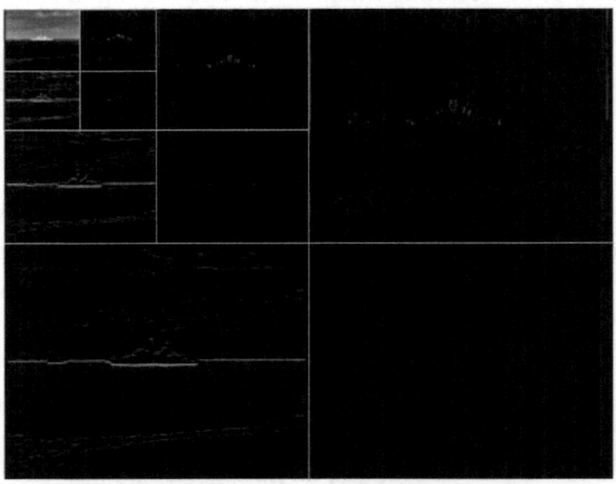

(b) An image of wavelet decomposition into three levels

Fig. 3. 2D Discrete Wavelet decomposition into three levels

2.2 Edge Detection Based on 2D Discrete Wavelet Transform

Generally, in IR images of sea environment, object regions such as ships and buoys have brighter intensity values than background region such as sea surface and sky. Because object regions seem to be irregular texture regions that have various intensity values, horizontal and vertical edges can be appeared complexly. Horizontal edges are mostly appeared in background regions. In order to effectively detect edges, we perform 2D Discrete Wavelet Transform to the noise removed image.

2D Discrete Wavelet Transform (2D DWT) [9,10] is used in image processing as a powerful tool solving to image analysis, denoising, image segmentation and other. 2D DWT can be applied as a convolution of a selected wavelet function with an original image or it can be seen as a set of two matrices of filters, row and column one. Using a separable property of DWT, the first part of decomposition consists of an application of row filters to the original image. The column filters are used for further processing of image resulting from the first step. This image decomposition [9] can by mathematically described by (5)

$$\mathbf{C} = \mathbf{X} \bullet \mathbf{I} \bullet \mathbf{Y} \tag{5}$$

where **C** is the final matrix of wavelet coefficients, **I** represents an original image, **X** is a matrix of row filters and **Y** is a matrix of column filters. In the first level of decomposition of 2D DWT, the image is separated into four parts. Each of them has a quarter size of the original image [10]. They are called approximation coefficients (LowLow or LL), horizontal (LowHigh or LH), vertical (HighLow or HL) and detail coefficients (HighHigh or HH) [10,11]. Fig.3 shows Discrete Wavelet decomposition into three levels. As shown in Fig.3(b), we can find out that horizontal and vertical edges are appeared in object regions, simultaneously.

2.3 Saliency Map Generation

In this paper, in order to effectively generate a saliency map, we firstly create three combined maps by combining horizontal and vertical edge images in each level. Then the three combined maps are integrated into a saliency map. However, as shown in Fig. 3(b), because the horizontal edges do not correctly correspond to the vertical ones, it is difficult to detect the strong edge region for both directions. Therefore, before combining the two edge images in each level, we perform the Gaussian blurring to each edge image by using (4).

$$\begin{aligned} &CE = G(\tilde{f}_{vertical}) * G(\tilde{f}_{horizontal}) \\ &G(\bullet): Gaussian \;\; blurring \end{aligned} \tag{6}$$

where $\tilde{f}_{vertical}$ and $\tilde{f}_{horizontal}$ are vertical and horizontal edge image in each level. Fig. 4 shows the three combined images by using (6). Finally, a saliency map is created by (7).

$$SM = CE_{Level1} + CE_{Level2} + CE_{Level3} \tag{7}$$

Fig. 6 shows three combined maps in each level and a saliency map.

2.4 Object Region Extraction

In this paper, in order to effectively extract object regions, we use a binarization technique. The representative method for image binarization is Otsu method[12]. However, in Otsu method, as the size of an image is getting larger, the amount of calculation

(a) An enhanced image

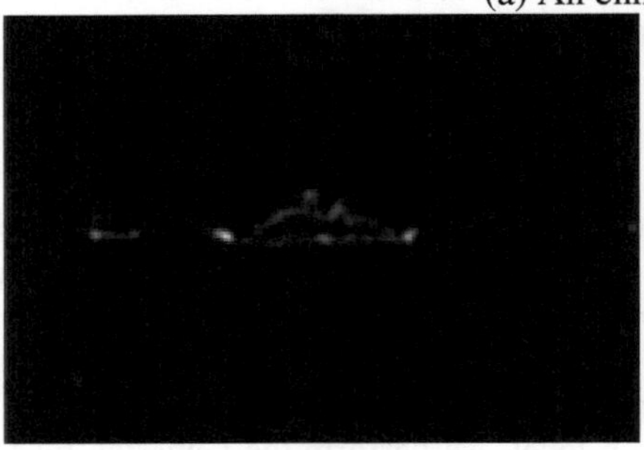

(b) Combined map of 1st level

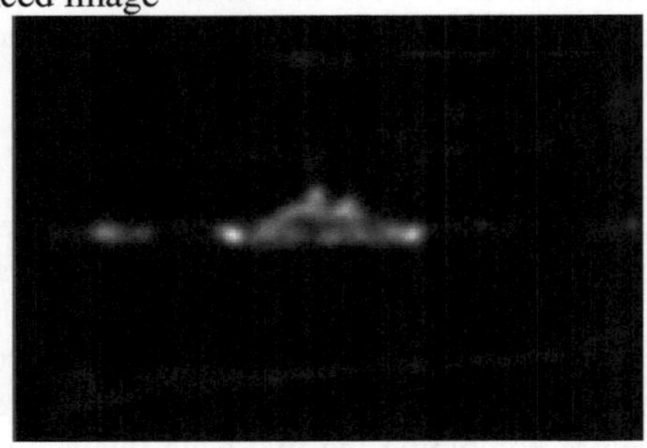

(c) Combined map of 2nd level

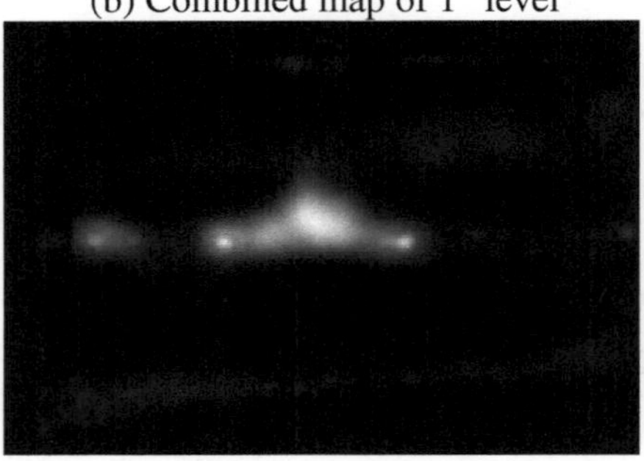

(d) Combined map of 3rd level

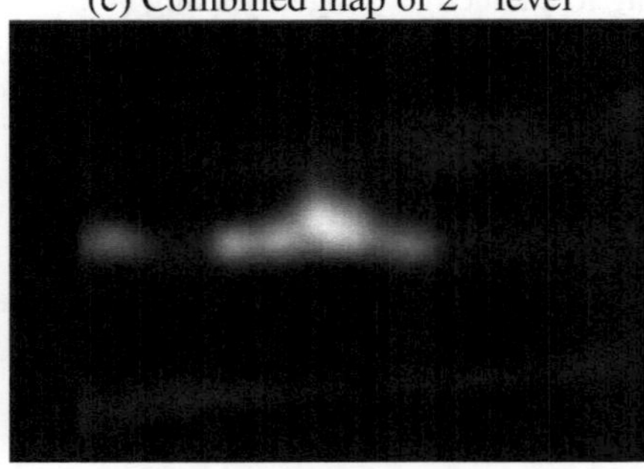

(e) Saliency map

Fig. 4. Result images of three combined maps and a saliency map

is increasing. That is why this method is not suitable for real-time processing. Therefore, we use not only the statistical method like Otsu method to extract the object regions from IR images with various qualities but also the method that calculation complexity is relatively low. The threshold for image binarization can be calculated by (8). *mean*(·) and *std*(·) are the mean and the standard deviation of the strength of edges in the image, respectively. The weight α is decided 1.5 through the experiments. Consequently, the regions that have the edge strength over the threshold are extracted as object regions. Fig. 5 shows the object regions extracted from an input image by using (8).

$$Obj_{candidate} = \begin{cases} 255 & CE(x,y) \geq T \\ 0 & CE(x,y) < T \end{cases}$$

$$T = mean(CE) + \alpha \cdot std(CE) \tag{8}$$

(a) Binaryzation result (b) Final result

Fig. 5. Extraction result of object regions

3 Experimental Results

In the experiments, we evaluate the performance of the proposed method using various IR images that contain ships and buoys in sea environment. Fig. 6 shows the experimental results of extracting the object regions from the input images with noises in the backgrounds. The proposed method effectively removes noisy regions and correctly detects the object regions by extracting the edge regions and performing morphological operations. The case of Fig. 6(a) and 6(b) is that a ship is located in the sea area. The case of Fig. 6(c) and 6(d) is that a ship is located on the horizon. In Fig. 6(e) and 6(f), objects with various sizes are shown. The experimental results show that the proposed method can detect the object regions in various sea environments.

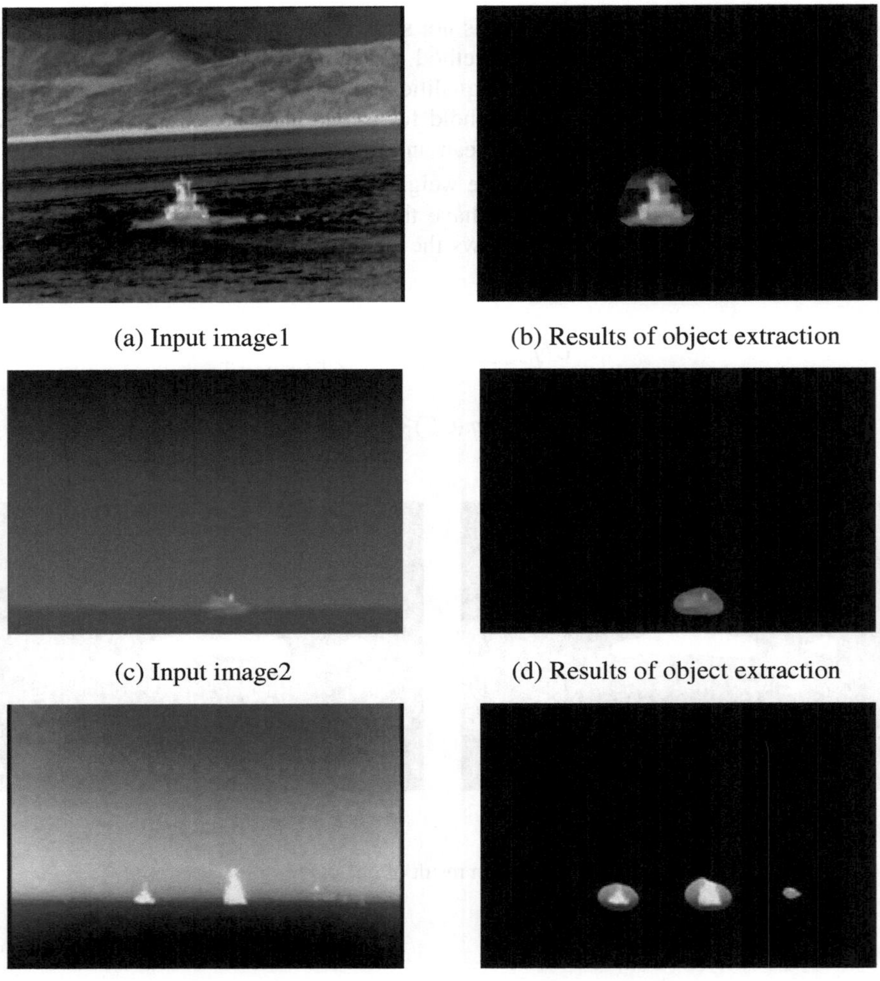

(a) Input image1 (b) Results of object extraction

(c) Input image2 (d) Results of object extraction

(e) Input images (f) Results of object extraction

Fig. 6. Extraction of object regions in sea environment

4 Conclusions

In this paper, we proposed the algorithm for detecting the objects in sea environment by using the directions of edges. To this end, we firstly remove the noises such as sea glints and complex clutters. Subsequently, we extracted edges by 2D DWT and generated the edge map. In the next step, a saliency map was created. Finally, we extracted the final object regions by performing a thresholding technique. The experimental results show that the performance of the proposed method is good and the proposed method correctly detected the object with various environments. We expect that the necessary information for navigation obtained by the proposed method will be supplied together with other equipments.

Acknowledgement. This research was a part of the project titled 'Development of the Intelligent Technology for Marine Accident Protection & Salvage', funded by the Ministry of Oceans and Fisheries, Korean.

References

1. Kim, D.J., Kwak, S.Y.: Evaluation of Human Factors in Ship Accedents in the Domestic Sea. Journal of the Ergonomics Society of Korea 30(1), 87–98 (2011)
2. FaulKemer, D.: Shipping Safety. Ingenia (2003)
3. Toffoli, A., Lefevra, J.M., Bitner-Gregersen, E., Monbaliu, J.: Toward the Identification of Warning Criteria: Analysis of a Ship Accident Database. Journal of Applied Ocean Research 27, 281–291 (2005)
4. Jeong, J., Park, G.-K.: Object Detection Algorithm Using Edge Information on the Sea Environment. Journal of the Korea Society of Computer and Information 16(9), 69–76 (2011)
5. Gaarder, S., Rongstad, K., Olofsson, M.: Image of human elements in marine risk management. Guedes Soares C., Advances in safety and reliability, pp. 857–898. Pergarmon (1997)
6. Vicker, V.E.: Plateu equalization algorithm for real -time display of high-quality infrared imagery. Optical Engineering 35(7), 1921–1926 (1996)
7. Brustrom, K., et al.: Object detection in cluttered infrared images. Optical Engineering 42(2), 388–399 (2003)
8. Gonzalez, R.C., Woods, R.E.: Digital image precessing, 2nd edn. Prentice Hall (2001)
9. Barni, M.: Document and Image Compression. CRS Press, Taylor and Francis Group (2006)
10. Nguyen, T.Q.: A tutorial on ¯lter banks and wavelets. University of Wisconsin, ECE Department (June 1995)
11. Bovik, A.: The Essentuial Guide to Image Processing, 2nd edn. Elsevier, Inc. (2009)
12. Ostu, N.: A threshold selection method from gray-level histograms. IEEE Transactions on Systems, Man, Cybernetics SMC-9, 62–66 (1979)

and [illegible] [illegible]

Detection Method of Radio Frequency Interference Using Raw Measurement of Multi-GNSS Receivers

Ki-Yeol Seo, Won-Seok Jang, and Young-Ki Kim

GNSS Research Center, Korea Institute of Ocean Science & Technology, Korea
kyseo@kiost.ac

Abstract. This paper focuses on the detection method of radio frequency interference (RFI) and the system using multi-GNSS receivers at current DGPS reference station. If the DGPS reference station is affected by radio frequency interference (RFI), it will not be able to perform the functionalities of DGNSS RSIM, as well as affect the quality of the pseudorange because the reference antenna cannot receive the GNSS signals correctly. Therefore, using multi-DGNSS receivers, it proposes the simple method for detecting the RFI at current DGNSS station, and shows the test results. First of all, this paper introduces the configuration and functionalities of DGNSS RSIM, and presents the limitation of integrity monitoring function, when the RFI occurred in the current DGNSS station. With the developed DGNSS software RSIM system interfaced with multi-GNSS receivers, it analyzes the raw measurement output of the receivers, and then proposes the detection method of the RFI, and finally, it summarizes the test results using a RFI simulation system. Based on the test result, it was able to verify the feasibility of RFI detection using the proposed method.

Keywords: Radio Frequency Interference (RFI), Differential GNSS, RSIM, RTCM, RFI detection method.

1 Introduction

Radio frequency interference (RFI) has a significant influence on the tracking of GNSS satellite signals by a GNSS receiver. This can range from a degradation of the performance to a total prohibition of satellite signal acquisition and tracking. Any signals transmitted in, or near the GNSS frequency bands will interfere with the reception of the GNSS signals. Different types of interference are: unintentional interference of other systems using the same frequency bands, intentional interference also known as jamming and naturally occurring interference. Several effects of interference can be detected on the output of a GNSS receiver [1]. In order to emphasize the relevance of a vulnerability assessment on navigation applications, we will consider the effects of RFI in more detail differential GNSS reference stations and integrity monitors (RSIM). There are several ways to limit the influence of RFI on receiver performance. It is the purpose of this study to assess the possibilities of detecting RFI

S. Kim et al. (eds.), *Soft Computing in Intelligent Control*,
Advances in Intelligent Systems and Computing 272,
DOI: 10.1007/978-3-319-05570-1_6, © Springer International Publishing Switzerland 2014

without using dedicated hardware. This paper focuses on the influence of radio frequency interference (RFI) on GNSS receivers at a DGNSS reference station, when RFI occurs in the current maritime DGNSS. With the developed software-based RSIM system interfaced with GNSS receivers, it analyzes the raw measurement output of the receivers. First of all, it describes the configuration of DGNSS system and RTCM RSIM messages, and then classifies the specific parameters of the messages that are affected by RFI, and finally, it summarizes the output results by RFI type. Based on this result, it will be able to offer advantage as a significant process for setting the threshold for RFI detection and identification at the DGNSS RSIM system.

2 DGNSS RSIM

Fig. 1 presents the recommended functional partitioning of the various elements that constitute a DGNSS service, based on software functionality [2]. The software-based reference station (S/W RS) uses a reference receiver, and a minimum-shift keying (MSK) modulator to generate the RTCM messages for broadcasting.

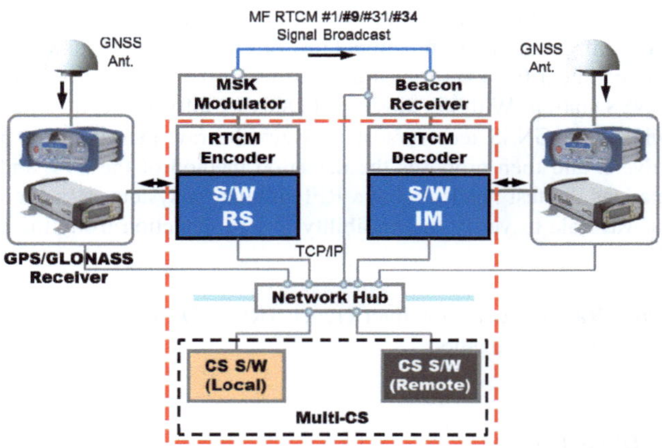

Fig. 1. System configuration of software-based DGNSS RSIM

The software-based integrity monitor (S/W IM) receives the RS broadcast, and verifies that the information is within tolerance. The IM routinely provides positive system feedback to the RS, to indicate normal, monitored operation. During out-of-tolerance conditions, the IM generates alarms. The most important alarms are the position, and pseudorange correction (PRC) alarms. These require a simple radial error check of the DGNSS navigational solution against a known position, and an analysis of the accuracy of individual ranges from the satellites to the known fixed position of the IM. The CS manages the service, by setting or resetting the equipment parameters, including data recording intervals, modes of operation, alarm thresholds and intervals, radio-beacon almanac information, and the RTCM message broadcast schedule [3]. In order to monitor the integrity, the DGNSS system checks the maximum pseudorange corrections (PRCs) and range rate corrections (RRCs), RSIM feedback message, UDRE, minimum number of satellites being tracked, correction age, pseudorange and range rate (PR/RR) residuals, absolute position error, DOP, message error ratio (MER), and signal strength (SS). Based on the RTCM RSIM document [4], the performance requirements related to the integrity monitor are summarized, as shown in Table 1 [5].

Table 1. Parameters for integrity monitor

Parameters	Threshold	Systems
Minimum number of satellites	4	RS/IM
Maximum PRC	100m	RS
Maximum RRC	4m/s	RS
IM feedback message	12s	RS
Correction age	30s	IM
PR residual	5m	IM
RR residual	0.5m/s	IM
Absolute position error	10m	IM

A set of RSIM messages is defined, in order to standardize equipment interface parameters, and to facilitate the flow of information between the various equipment, such as reference station, integrity monitor, transmitter/RSIM interface module, and control station for DGNSS service, regardless of the manufacturer. There are 40 message types for RTCM SC-104 RSIM version 1.2 [4], and the RSIM Messages affected by RFI, and their parameters, are listed in Table 2.

Table 2. Parameters of RSIM messages

RSIM #	Descriptions	Related Parameters
7	GPS Receiver Satellite Status	C/No
12	Reference Station Alarms	Insufficient Satellite Alarm, High PRC/RRC Alarm
13	Reference Station Correction Data	PRC/RRC
17	Integrity Monitor Alarms	Correction Age, Position Error, DOP, PR/RR Residual Alarms
18	Integrity Monitor DGPS Status	Position Error, Quantity of PRNs, DOPs
19	Integrity Monitor Correction Data	PR/RR Residuals

3 RFI Detection Method

In order to analyze the effects of RFI on GNSS receiver output, the quality monitoring of receiver's raw measurement output was identified. The following parameters are used to determine the RFI identification. The first step1 is to analyze the raw measurement from the reference receivers, and then set the 4 parameters such as SVs, PRC/RRC, C/No, position error, as shown in Fig. 2.

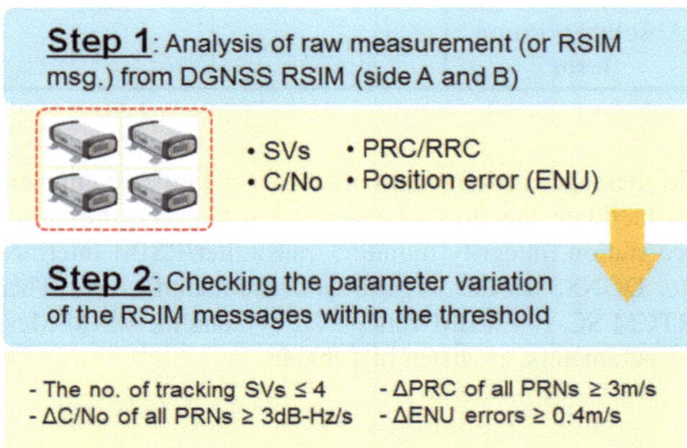

Fig. 2. RFI detection process

Fig. 3 presents a flowchart for RFI detection. We designed a simple algorithm for RFI detection using the parameters and the threshold values.

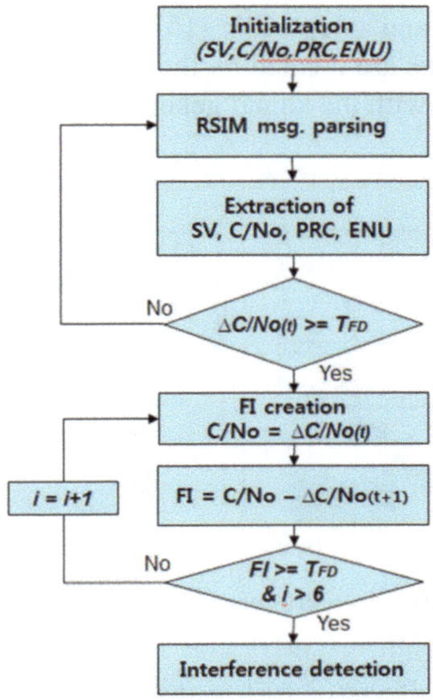

Fig. 3. Flowchart for interference detection

4 Experimental Setup

Table 3 presents the test environment of the radio frequency interference (RFI) with a GNSS simulator, a interference signal generator, and reference receivers. In order to generate the GNSS signal and interference scenario, the systems consists of an operation program (SimGEN), GNSS signal generator (GSS8000), signal generator for RFI generation, and user receiver for raw measurement output.

Table 3. RFI test environment

	System	Description
GNSS scenario creation	SimGen S/W	GPS/GLONASS operation scenario
GNSS signal generator	Spirent GSS8000 12CH	GPS/GLONASS L1/L2 signal generation
RFI generator	Agilent MXG Vector Signal Generator	RFI scenario and signal generation
User receiver	Novatel DL-V3	Raw measurement output

The first phase is the implementation of a simulation environment for reconfiguration of the RFI status, and then is to develop the software platform for detecting the effect of RFI on DGNSS receivers. Fig. 4 shows the system connection for the GNSS simulation tests. The RFI signals used coherent continuous wave (CW) of GPS L1 frequency (1575.42MHz), and swept CW (1574.42 to 1576.42MHz) using a signal generator. The RFI scenarios are created by the SimGen software for operation of the GNSS simulator interfaced with the signal generator.

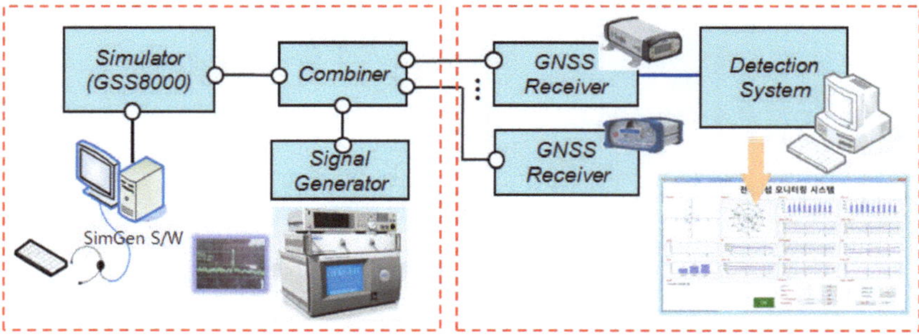

Fig. 4. Interference simulation setup

In order to analyze the effects of RFI at the reference station's receiver, we used off-the-shelf receivers. The analyzed parameters are the number of satellites (SVs), pseudorange corrections (PRCs), signal strength (C/No), and position accuracy. Fig. 5 and Fig. 6 present the number of satellites and PRCs, and Fig. 7 and Fig. 8 show the signal strength (C/No) of PRN# and the position accuracy in case of coherent CW interference, respectively.

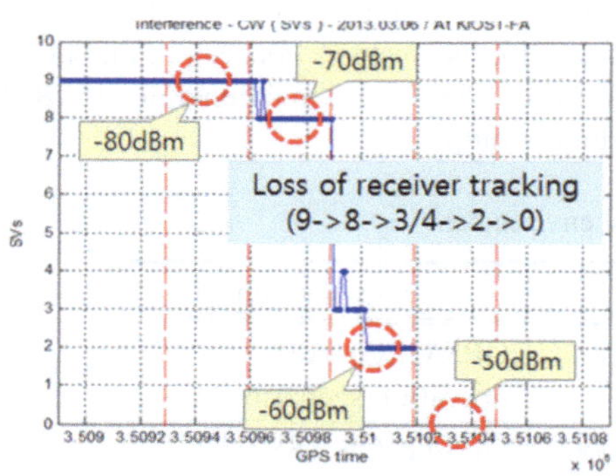

Fig. 5. The number of SVs – CW

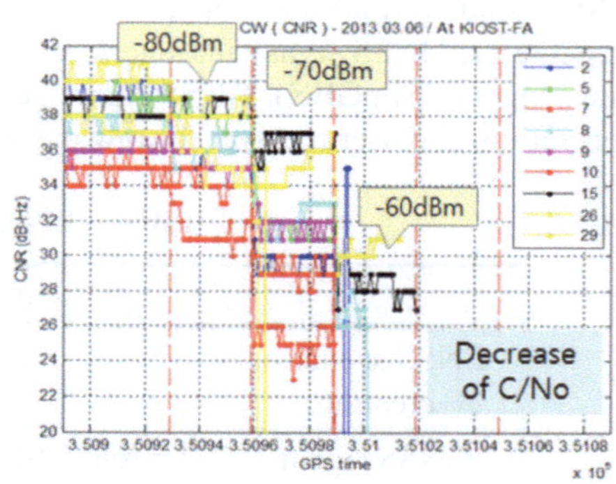

Fig. 6. C/No. of PRNs - CW

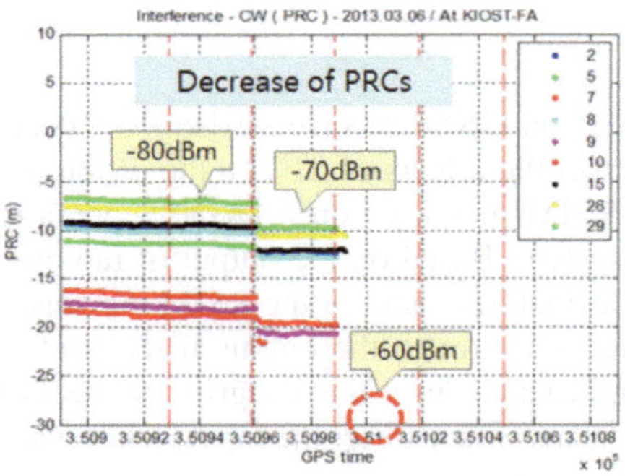

Fig. 7. Pseudorange corrections – CW

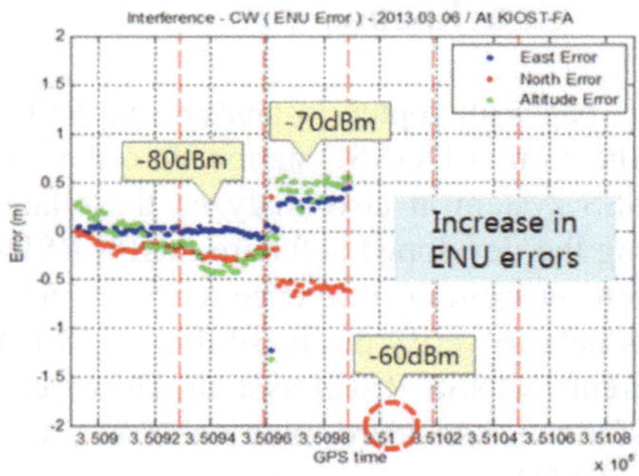

Fig. 8. Position accuracy (ENU) - CW

Based on the RFI simulation results, it was able to determine the following parameters and their threshold values:

- The number of tracking SVs ≤ 4
- ΔC/No of all PRNs $\geq$ 3dB-Hz/s
- ΔPRC of all PRNs $\geq$ 3m/s
- ΔENU errors $\geq$ 0.4m/s

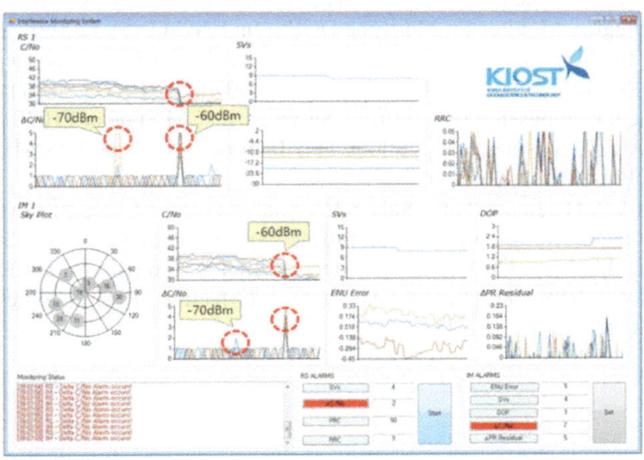

Fig. 9. Interference Monitoring Program

With respect to the analysis of the RFI effect at the DGNSS reference station, it analyzed the RSIM messages and the parameters. The time interval for analyzing the RSIM message was set to 1 second, and it was able to confirm the feasibility of the RFI detection through the test results.

5 Conclusions

This paper attempted to analyze the RFI effect on GNSS receivers when RFI occurs in the current DGNSS station. In order to detect the effects of RFI on the differential GNSS system, it also analyzed the related RSIM messages on a DGNSS operation using the developed software DGPS RSIM system. Based on the output of raw measurement from the reference receivers, and the analysis result of the RSIM message, it was able to verify the feasibility of RFI detection through the output analysis of the current DGNSS RSIM system. However, in order to meet the integrity of the IMO performance standard, such as TTA < 10s, it is necessary to set the logging interval of the RSIM message within 5 seconds. Therefore, if only a simple module for processing the RSIM message is installed at the reference station, it would be expected to enhance the integrity of the system.

Acknowledgement. This work was supported by the Ministry of Oceans and Fisheries (MOF) (No. PMS2840).

References

1. de Bakker, P.F.: Effects of Radio FrequencyInterference on GNSS Receiver Output. Master'sThesis, Faculty of Aerospace Engineering, Delf University of Technology
2. Seo, K.Y., Park, S.H., Jang, W.S., Suh, S.H.: Performance Analysis of Software Reference Station and Integrity Monitor for Maritime DGPS. In: ENC-GNSS (2009)
3. Radio Technical Commission for Maritime Services, RTCM Recommended Standards for Differential GNSS Version 2.3, RTCM Paper 136-2001/SC104-STD (2001)
4. Radio Technical Commission for Maritime Services, RTCM Standard 10401.2 for Differential NAVSTAR GPS Reference Stations and Integrity Monitors (RSIM), RTCM Paper 221-2006-SC104-STD (2006)
5. Seo, K.Y., Park, S.H., Jeong, H.C., Suh, S.H.: Analysis on the Limitation of Integrity Monitoring Functions for Maritime DGPS. In: Proceedings of International GNSS (2009)
6. Vogel, A.-L., Macabiau, C., Suard, N.: Effect of a GPS Anomaly on Different GNSS Receivers. In: ION GNSS 18th International Technical Meeting of the Satellite Division, Long Beach, CA, September 13-16 (2005)

Creating an Ontological Model for the Tax System in Kazakhstan

Altynbek Sharipbaev, Assel Omarbekova, and Shinar Turmaganbetova

Dept. of Theoretic Computer Science, L.N. Gumilyov Eurasian National University
2 Mirzoyana street, Astana, Republic of Kazakhstan

Abstract. Within the dissertation work the main concepts of ontological model were identified to organize the process of implementation of the expert system of the tax system in Kazakhstan. Thus an ontological model will be created to provide corresponding operating conditions of the system and implement a web-application that uses Semantic Web technologies and represents the interface of the expert system machine inference that works with these concepts.

Keywords: ontological model, Semantic Web, tax system.

1 Introduction

On the base of the L.N. Gumilyov Eurasian National University the researches on the development of the expert system, which will independently be an expert on an intellectual level are being conducted.

As a result every citizen will be able to get a response online from the tax system of Kazakhstan. This type of support is more economical to provide remotely, automatically, that is to use the formal knowledge of the expert. Knowledge, which provides technical support for a variety of subject areas, has a similar structure. To provide it, it is rational to use ontologies and other technologies of Semantic Web.

The user of this system, of advisory support, as a rule, does not know all of the cause-effect relationships of the system, and is able only to ask the expert questions.

As a result, when applying to the search system a user should be able to get a response relevant to the request, so the search should be semantically oriented. To do this, the appropriate means of finding information request is proposed to organize on the basis of an ontology containing the description of semantic of resources.

The task of the expert system of support is to identify the issue and give the user a full answer at incomplete input of information. And the intelligence of the system depends on how fast the system can respond to the question.

In computer science and information science, ontology formally represents knowledge as a set of concepts within a domain, and the relationships between pairs of concepts. It can be used to model and support a domain, reasoning the concepts.

In theory, ontology is a «formal, explicit specification of a shared conceptualization». Ontology provides a shared vocabulary, which can be used to model a domain, that is, the type of objects and/or concepts that exist, and their properties and relations [1].

S. Kim et al. (eds.), *Soft Computing in Intelligent Control*,
Advances in Intelligent Systems and Computing 272,
DOI: 10.1007/978-3-319-05570-1_7, © Springer International Publishing Switzerland 2014

Ontologies are the structural frameworks for organizing information and are used in artificial intelligence, the Semantic Web, systems engineering, software engineering, biomedical informatics, library science, enterprise bookmarking, and information architecture as a form of knowledge representation about the world or some part of it. The creation of domain ontologies is also fundamental to the definition and use of an enterprise architecture framework.

As a result of the dissertation research, a decision-making system which will be able to make decisions depending on a question asked by the user concerning tax system of Kazakhstan has to be created.

2 Experimental Part

As result of research an ontological model of tax system in Kazakhstan was created (Figure 1).

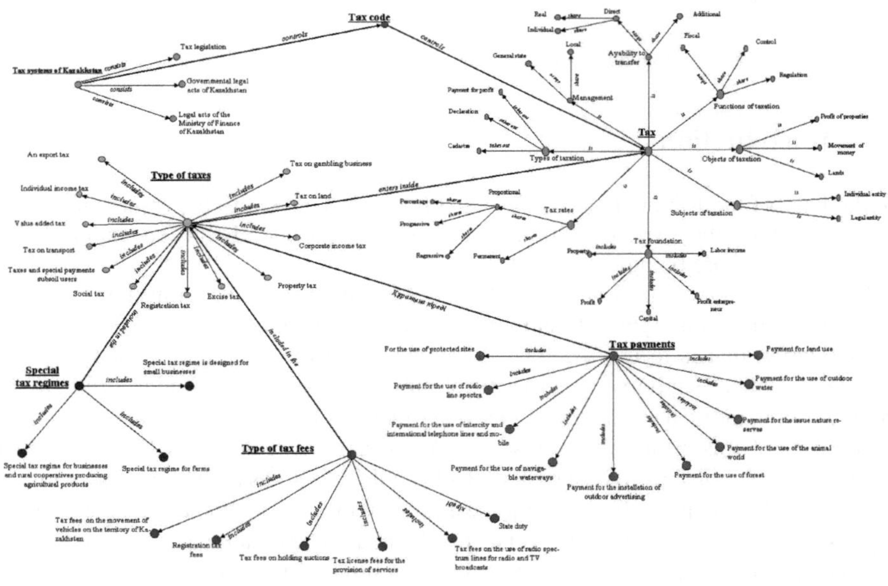

Fig. 1. Ontology of tax system in Kazakhstan

In artificial intelligence, an expert system is a computer system that emulates the decision-making ability of a human expert. Expert systems are designed to solve complex problems by reasoning about knowledge, like an expert, and not by following the procedure of a developer as is the case in conventional programming.

An expert system has a unique structure, different from traditional computer programming. It is divided into two parts, one fixed, independent of the expert system: the inference engine, and one variable: the knowledge base. To run an expert system, the engine reasons about the knowledge base like a human [2].

As a consequence of these objective circumstances it is highly desirable that the information and knowledge have been described and structured in such a way that the recipient (user) was able to understand the text and the context (the meaning) of the message. Ideally, the message (knowledge entity) should be structured in such a way that the computer, not just the educated person was able to "understand" it. The word «understand» here means that the computer is able to process the document (knowledge entity) through the use of rules known to him by some logical language, and will be able to bring new facts and knowledge of the document (Figure 2).

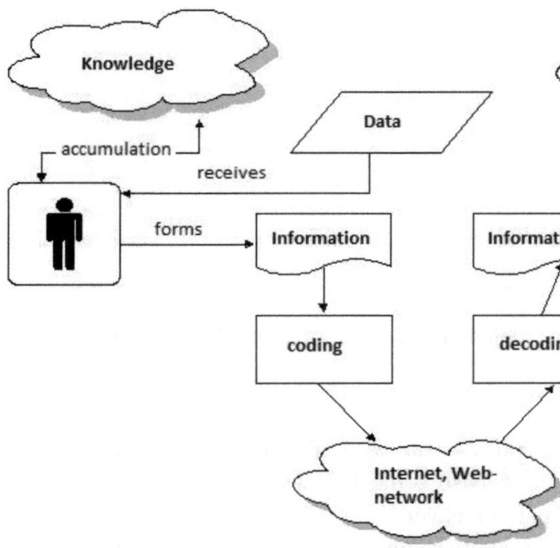

Fig. 2. Scheme of functioning of the system shows how many figures interact with each other in the work processing

Thus systems could be doing follow activities:

- formation and fixation of the total, shared by all the experts domain knowledge (software)
- clear conceptualization of software that allows you to describe the semantics of the data,
- opportunity to re-using knowledge,
- integration and the ability to share heterogeneous data and knowledge in a single system,
- provide a better understanding of the subject area of system users.

The system in response to a specific question collects your reply automatically or builds algorithm on the basis of all the necessary regulations related to the question of the particular user (Figure 3).

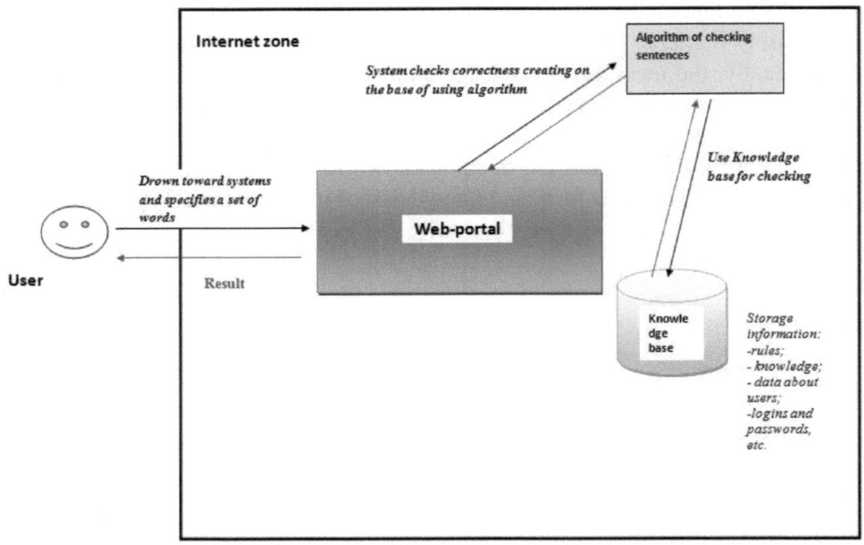

Fig. 3. Main elements of system, which transfer information to each other till they get necessary result

The user asks the system a question – the System defines category of a tax – the System addresses to the Knowledge base – the Knowledge base processes inquiry - the Knowledge base gives the created answer.

The system comprises a knowledge base and three subsystems: subsystem control, a user interface, subsystem inference. All subsystems are separate software modules that work as part of the overall executable program.

The system is implemented in a multi-menu-based interface. During the dialogue, the system displays various kinds of clues. In the calculations the initial data is entered from the keyboard. The calculation results are displayed on the screen as well as to the printing device (by the user). The whole system is designed so that the user can work with it without any special training.

Thus, we have identified the main elements in the created ontological model of the tax system in Kazakhstan. There are external elements, which are associated with the new system, further highlighted the internal elements of the system. Also highlighted are the main factors that affect the change of the system in a temporary space (Figure 4).

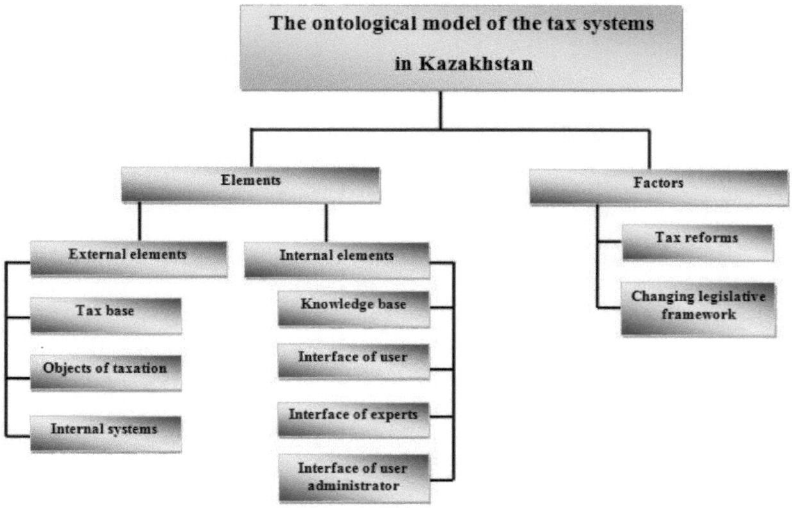

Fig. 4. The ontological model of the tax system in Kazakhstan that consists of elements (internal and external) and factors that influence the system

3 Conclusion

As result of research we have created ontology model of the tax system in Kazakhstan, which will improve the existing tax system in our state and to develop the expert systems that could link citizens with government.

As a result of introduction of the system the following tasks will be realized:

- automation of work process with citizens.
- transparency of the system
- lack of the human factor

Acknowledgements. The authors express huge gratitude to L.N. Gumilyov Eurasian National University in Kazakhstan and Tax committee of Kazakhstan, who made a tremendous contribution in promotion of the research.

References

1. Berners-Lee, T., Hendler, J., Lassila, O.: The Semantic Web. Scientific American (May 17, 2001)
2. Stewart, S.: Tax-based expert systems: a first principles approach. Expert Systems. In: O'Leary, D.E., Watkins, P.R. (eds.) Expert Systems in Finance. Elsevier Science Publishers B.V. (1992)

2030 Toy Web of the Future

Boyoung Kang, Jeonghwan Heo, Heidi Hye Seung Choi, and Kwang H. Lee

Department of the Graduate Program of Future Strategy, KAIST,
291 Daehak-ro, Yuseong-gu, Daejeon, Republic of Korea
{careersmart,axismundi}@naver.com,
hhsc87@gmail.com,
khlee@kaist.ac.kr

Abstract. Toy web is defined as web-based interactive contents for children and teen agers to improve their cognitive ability. Its market is just emerging now and we would like foresight how its user interface in 2030 will look like. In order to foresight we have to identify key drivers and related variables. We used STEPPER for this, and 3 Layer timed-SWOT(three present and future SWOT(strength, weakness, opportunity, threat) analyses based on three key drivers; technology, education, policies). The final step for the future foresight is to drive the future strategies and foresight integration, thereby creating toy web foresight scenarios. Our overall foresight process is summarized as follows;

Keywords: Toy Web, STEPPER, Timed-SWOT, 3 Layered Timed-SWOT, Cognitive Science, Metaverse, Universal Browser, Augmented Reality.

1 Introduction

The overview of our entire process of foresight can be exhibited as follows;

First, we wanted to define and organize the issues in the team meeting. We chose the topic at the future workshop and met a few times and correspond with tens of e-mails and SNS. Second, we wanted to identify key drivers and related variables. We used STEPPER for this. Third, we wanted to figure out the relationship between the core element of Toy web and the key variables. We drew the relation map for this. Fourth, we wanted to analyze the current situation of Toy web environment and the gap between the desirable and probable future of 2030 in terms of three key drivers; technology, education, policies). We used 3 Layered timed-SWOT(three present and future SWOT(strength, weakness, opportunity, threat) for this and drove the future strategies. The final step for the future foresight is to integrate all the four steps and to create toy web foresight scenarios.

S. Kim et al. (eds.), *Soft Computing in Intelligent Control*,
Advances in Intelligent Systems and Computing 272,
DOI: 10.1007/978-3-319-05570-1_8, © Springer International Publishing Switzerland 2014

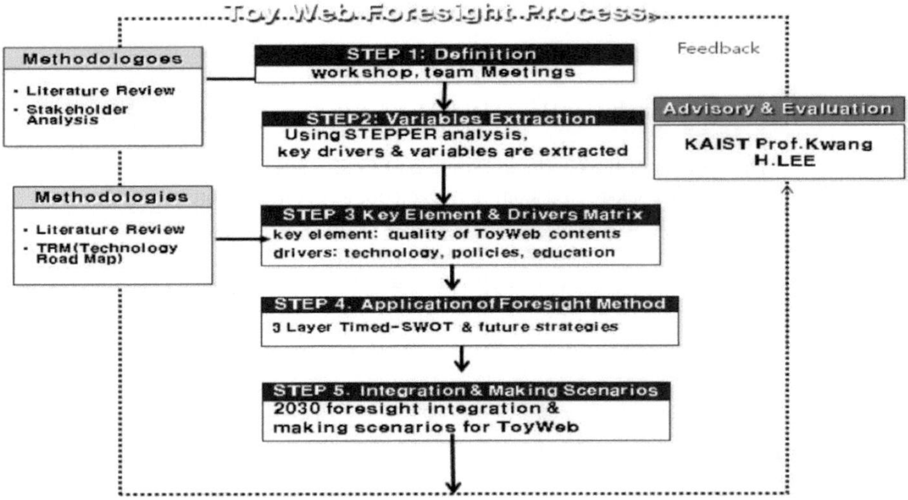

1.1 STEP 1. Project Definition

Toy web is defined as web-based interactive contents for children and teen agers to improve their cognitive ability. The core technology of Toy web is the multi-modal interface and interaction. Cognitive ability technology has been the meaningful convergence product of the four most important technologies, so-called NBIC(NT, BT, IT, CS(cognitive science)) as the U.S : National Science Foundation also mentioned in 2011.

(Picture Source: National Science Foundation and National Nanotechnology Initiative,2011)

The project is implemented as the academic practice for the structured future foresight method and the visualization of the future images of the internet and learning environment to give the stakeholders of the related industry. Through the entire process the purposes and the specific goals to try to drive the key element and drivers to realized the desirable 2030 Toy web environment. The gap analysis between the present and the future in terms of the important drivers such as technology and the possible strategies should be the final output of this research.

1.2 STEP 2. Finding Drivers and Variables

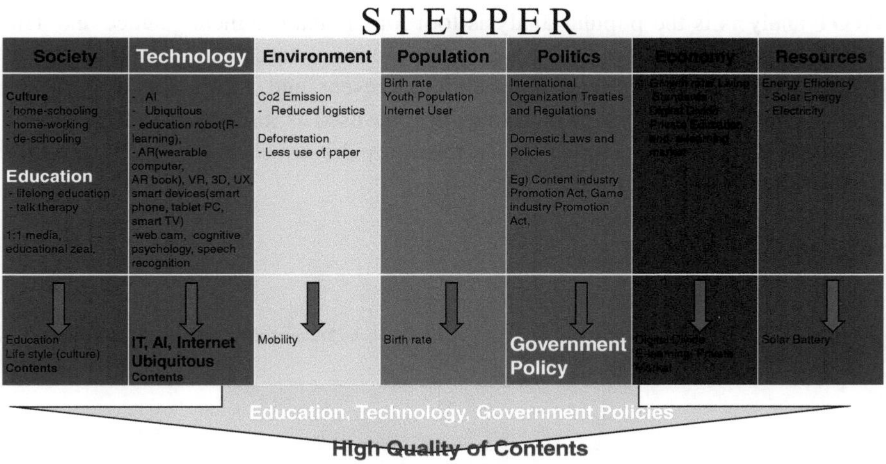

STEPPER is the framework for finding the key drivers of the defined subject. It is the scientific screening instrument like the funnel of the all the possible macro factors affecting the dependent variable which we are interested in. Using the STEPPER funnel, we finally drove the three key drivers; i) education infrastructure, ii) government policies, iii) technology, and came to a conclusion that the core element behind all these three is the quality of the Toy web contents.

1.3 Step 3. Key Element and Three Drivers Relation Map

We drove the relation map for the key drivers and relation variables. Among these independent variable, the technology is the most important driver positively influencing the high quality of contents of Toy web.

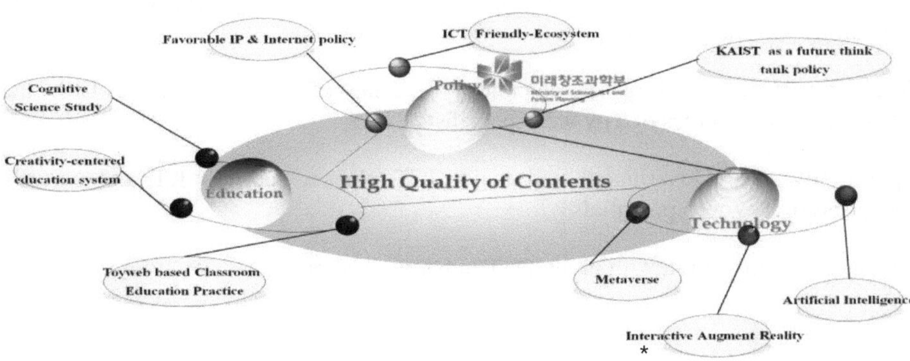

- Independent Variable: Policy, Education, Technology
  Dependent Variable: High Quality of Contents

*Metaverse: meta+universe, more advanced technology than virtual cyber space like second life

1.4 STEP4. 3 Layered Timed-SWOT

SWOT analysis is the popular analysis tool for the management science and Timed-SWOT by professor, Kwang H.LEE is the modified foresight method to drive the strategies to shorten the gap between the present and future and to realize the desirable future. The version of framework we used is another modified version of Timed-SWOT to analyze based on the three key drivers each and finally produced 3 layered Timed-SWOT.

Timed SWOT 1: Technology

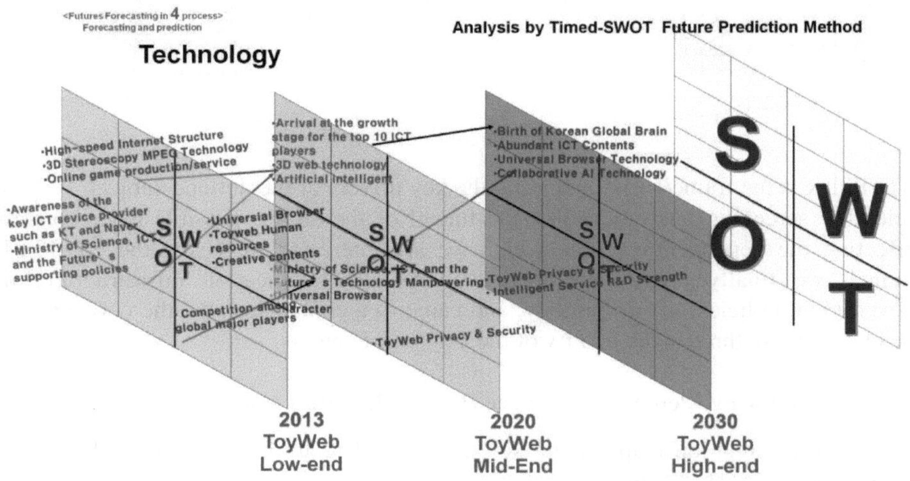

Strength: Korea's Telecommunication infrastructure, 3D Stereoscopy MPEG technology, Online game production/service technology

Weakness: *Universal browser technology time lag-4.5 years behind the U.S., the lack of human resources of Toy web specialists

Opportunity: The Government and major ICT players such as KT and Naver's awareness of Toy web potentials

Threat: The competition with global ICT industry players

*Universal browser is a fully functioning web browser tool giving you access to Internet through any devices any time.

Future Technology Strategies	Universal Browser, Augmented Reality, Collaborative Artificial Intelligence, ICT big 10 enterprises enter into Toy web market

Timed SWOT 2: Education Infrastructure

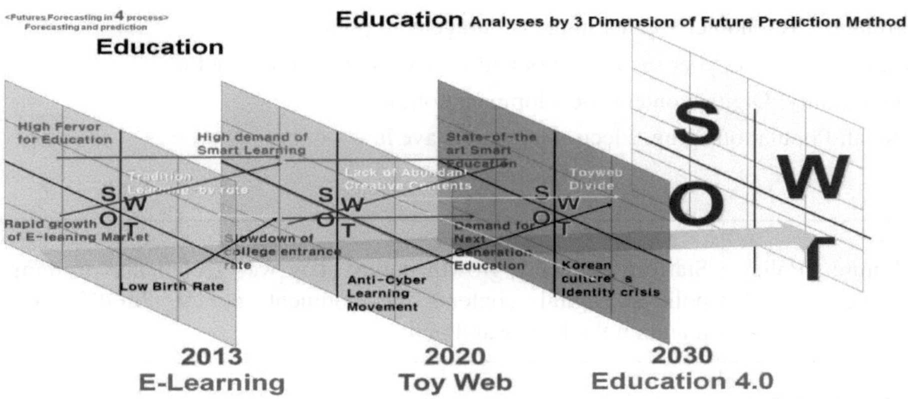

Strength: High national enthusiasm for the education,

Weakness: Severe competition for the college entrance, teaching by rote

Opportunity: e-learning market growth

Threat: Low birth rate

Future Education Strategies	Seminar and discussion class methods, Creativity focused education, Digital contents development, Cognitive ability development learning class using Toy web

Timed SWOT 3: Policy

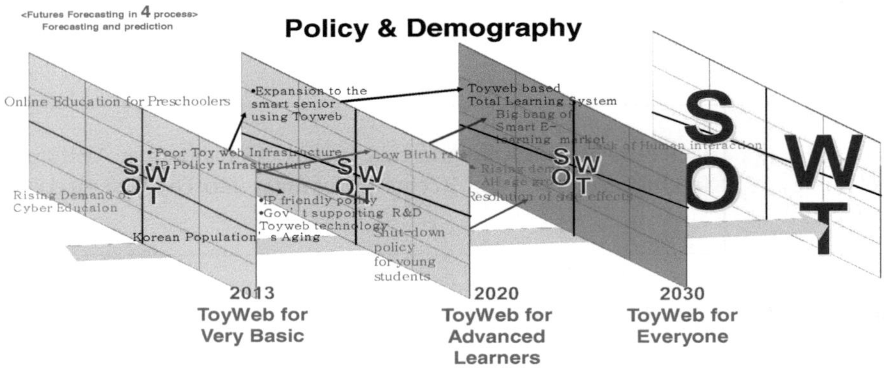

Strength: Preschooler's E-learning encouragement policy

Weakness: Strict cyber security , Poor IP Policy and law infrastructure

Opportunity: Digital contents development policy

Threat: Population aging, Electromagnetic wave law.

Future Policy strategies	Start-up company growth policy, Toy web technicians training policy, Digital contents development policy. Modify and maintain the IP law and policy

1.5 Step 5. Integration and Scenario making

The whole concept of Toy web is that if anyone wants to learn, they can learn anytime with any device, and learning should be fun. We made two scenarios, one for children, one for adults.

Scenario1: 2030 Toy Web for Children

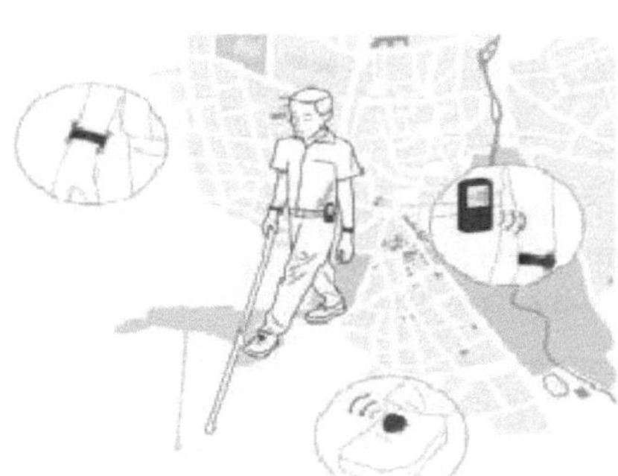

(Picture Source: ETRI Cognitive ability development convergence technology, June 2011)

In 2030 children in Korea have never been happier. They are very creative and efficient learners. Visually handicapped, Suncheol does not go to the school for the blind; he goes to the regular school. He is very happy with going to school using the haptic guiding devices listening to the audio information.

Scenario2: 2030 Toy Web for Adults

It it's not fun, it's not a car. Every daily activity is gamificated. Mr. Toyodasi is learning while he is smart auto-driving with hologram tutor until he arrives at the destination.

(Toyota's Fun VII (Vehicle Interactive Internet)

2 Conclusions

This research was to visualize the future of Toy web using the future foresight method. Driving the key drivers, analyzing SWOT, and visualizing the future can lead us to build the strategies and action plans for 2020 and for 2030 to fill the gap in terms of technology, education, and policies between the present and the desirable future and finally enable us to realize our vision.

The limit of this research is lack of quantitative analysis and the weight should be laid differently on the each variable according to the importance of effects to the dependent variable-the quality of Toy web contents. In spite of that, the most important key driver seems to be the metaverse related technology for the state-of the art Toy web; universal browser, augmented reality, collaborative artificial Intelligence. More specific and further technology strategies based on the TRM(Technology Road Map) analysis we should cooperate with internet and IT professionals and the causal relationship among the variables using system dynamics programs will make up for this research in time.

References

White Papers & Journals

- National Science Foundation and National Nanotechnology Initiative (2011)
- ETRI vol. 26-3 Cognitive ability development convergence technology (June 2011)

Websites

- Interaction Design Foundation, `http://www.interaction-design.org/encyclopedia/human_computer_interaction_hci.html`
- Futuristic News, `http://futuristicnews.com/`
- The Verge, `http://www.theverge.com/2011/11/28/2594660/toyota-fun-vii-concept-display-car`

Cooperation Level Estimation of Pair Work Using Top-view Image

Katsuya Sakaguchi and Kazutaka Shimada

Kyushu Institute of Technology, 680-4 Iizuka Fukuoka, Japan
{k_sakaguchi,shimada}@pluto.ai.kyutech.ac.jp

Abstract. To understand an interaction among persons is one of the most important tasks in artificial intelligence. In this paper, we propose a method for estimating a cooperation level in pair work. The task is a cooperation work that take place in front of a whiteboard by two persons. The goal of our study is to provide the cooperation level that is estimated by features extracted from images for teachers. The result of this study is useful for education support systems and problem based learning. We extract the standing location, operation ratio and head direction of each person from an overhead camera. We apply the features to two machine learning approaches: AdaBooost and multiple linear regression. We obtained 77.5% as the accuracy by the AdaBoost and 0.649 as the adjusted R^2 by the regression.

Keywords: Interaction analysis, Cooperation Level, Pair Work, Top-view Image.

1 Introduction

To understand an interaction among persons is one of the most important research tasks in artificial intelligence. We have proposed methods for understanding interactions in a conversation with spontaneous utterances [12,14,19]. In these studies, we focus on linguistic, phonetic and prosodic features. Utilizing information extracted from images is, however, necessary for understanding an interaction. Image data contains much information that linguistic information does not contain. Vargas [16] has reported that posture and gaze information are effective elements for estimating speaker's mind as the regulator that is actions such as a nod and a prompt of the next utterance. Mahmoud et al. [8] have reported an analysis of hand-over-face gestures for automatic inference of cognitive mental states. Kumano et al. [7] have analyzed how empathy and antipathy, aroused between people while interacting in face-to-face conversation, are perceived by external observers. In this paper, we also analyze an interaction with two persons by using features captured from images.

Recently, faculty development, which is to improve skills and knowledge about teaching ability, has been more important. For the purpose, Yamane et al. [18] have proposed a method to detect an interaction between a lecturer and learners. Furthermore, problem-based learning (PBL) is the more recent and highly

S. Kim et al. (eds.), *Soft Computing in Intelligent Control*,
Advances in Intelligent Systems and Computing 272,
DOI: 10.1007/978-3-319-05570-1_9, © Springer International Publishing Switzerland 2014

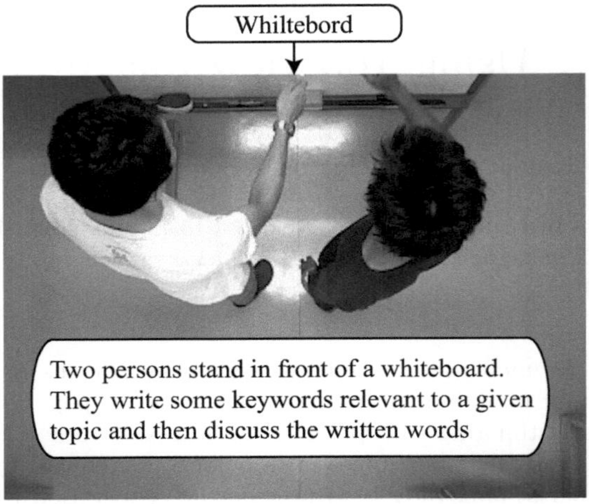

Fig. 1. An image of pair work from top-view

regarded. In PBL, students work in small collaborative groups and learn what they need to know in order to solve a problem [3]. The target of our research is the PBL environment.

In this paper, we propose a method for estimating a cooperation level in pair work. The cooperation level in this study is that "To what degree did a participant work in cooperation with a partner?"[1]. The task is a cooperation work that take place in front of a whiteboard by two persons. In our method, we focus on a top-view image for the estimation. Figure 1 shows an example of an image from an overhead camera. We extract the standing location, operation ratio and head direction of each person from the top-view image. We analyze the relation between the cooperation level and each feature. Then, we apply the features to two machine learning approaches: AdaBooost and multiple linear regression. The goal of our study is to provide the cooperation level estimated by the features to teachers in the PBL.

In the next section, we explain related work. Next, we describe our method in Section 3. In the section, we discuss nonverbal information for the cooperation estimation first, and then describe features and classifiers based on machine learning approaches for the task. In Section 4, we discuss our experimental results in terms of the features and classifiers. Finally, we conclude our methods in Section 5.

2 Related Work

Many researchers have studied methods for estimating atmosphere and participant's mind in an interaction. Takashima et al. [15] have reported an analysis

[1] Note that we are not concerned with the quality of the output of each pair work in this paper.

of nonverbal cues and atmosphere in six person conversations. The nonverbal information was acquired with some sensor devices such as a 3D tracker, an acceleration sensor and a tablet device. Mota and Picard [9] have proposed a posture recognition method for a person seated in a chair. They used a leap chair with pressure sensors. In general, using particular devices is, however, costly and cumbersome for participants.

One solution for the issue is to utilize cameras for the extraction of nonverbal information. Nakamura et al. [10] have proposed a method for estimating learners' subjective impressions for an e-learning system. They used facial information, facial expressions, gaze and head poses from a stereo-camera. Jayagopi et al. [4] have proposed a method for mining and validating group speaking and gaze patterns. They captured images and speech from two web-cameras and a commercial array microphone. Grafsgaard et al. [2] have proposed a method for analyzing posture and affect for intelligent tutoring systems. They captured depth information with a Kinect. These studied, however, treated a e-learning system for one person and persons seated in chairs. Our research target is a PBL environment in which participants have actions.

In this paper, we use a depth camera for the extraction of nonverbal information. The task is pair work. A participant might be occluded by another participant if a camera is placed in front of participants. To solve this problem, we apply an overhead camera to our task. By using the overhead camera, the problem of occluded images is solved. We have reported the effectiveness of the use of the overhead camera for a person identification task [6,11] and a posture identification task [5]. In addition, the method with the overhead camera has the advantage that psychological resistance is reduced because the camera does not capture the face image. Furthermore, the restriction of the location of a camera is reduced because the camera does not need to capture the person's face.

3 Proposed Method

In our method, we use Microsoft Kinect[2] to capture pair work activities. The Kinect camera can handle depth information. We extract several features from the captured images. Finally, we estimate the cooperation level in each pair work on the basis of these features.

3.1 Nonverbal Information

In this paper, we focus on nonverbal information for estimating the cooperation level. The nonverbal information in this paper is divided into three categories: (1) standing location, (2) operation ratio and (3) head direction.

First, we record a pair work activity with Kinect. Our method detects head areas of each person by using depth information from the Kinect. Next it computes the centroid of each head area. We regard the centroid as the standing location of a person. Figure 2 shows an example of the process.

[2] `http://www.xbox.com/en-US/Kinect`

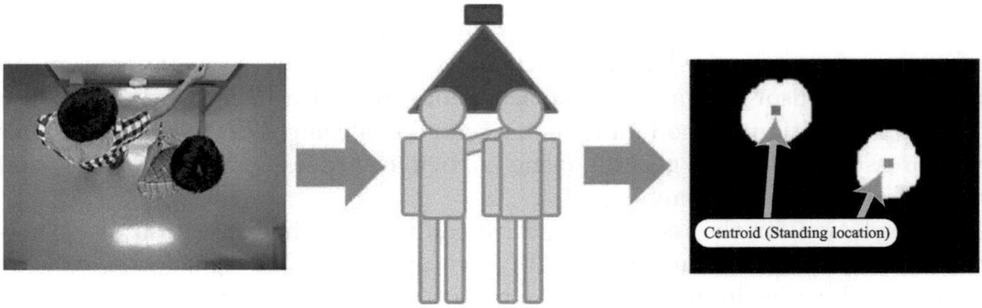

Fig. 2. The person area extraction and standing location

The second nonverbal information is an operation ratio of each person. Here the operation denotes a pointing gesture and a writing activity on the white-board. As a preliminary experiment, we generated a classifier for identifying the operations in a video. We used the locations of a pen tip and a fingertip by using depth information as the features for the classifier. In addition, we used the dimensions of person's area that are near the whiteboard and the distance between the centroids of the person area and head area as the features. Figure 3 shows an example of the process. We evaluated a machine learning technique using these feature with 300 images. The accuracy was 80%. We think that it is not sufficient to investigate the cooperation level in pair work although the accuracy is not low. Therefore, we prepare annotated data for the operation ratio. We manually annotate each image with three classes; pointing, writing and otherwise.

The third nonverbal information is a head direction. We divide an image to six directions. Figure 4 shows the six directions. We assign five directions to the front of the head and one direction to the backward. We also estimated the direction in each image automatically on the basis of the head shape. However, the accuracy was 47%[3]. Therefore, we also prepare manual annotated data for the head direction.

3.2 Features

We explained the nonverbal information that we use in the previous section. In this section, we describe features for the cooperation level estimation in detail. We introduce eight features to the estimation approach. These features are extracted from the output of the previous section.

Location deviation. We compute the standard deviation of the standing location. A large location deviation value denotes that the person is active during the pair work.

[3] The reasons why the accuracy was extremely low that were (1) a person sometimes stood outside of the camera range and (2) a head shape depended on the standing location.

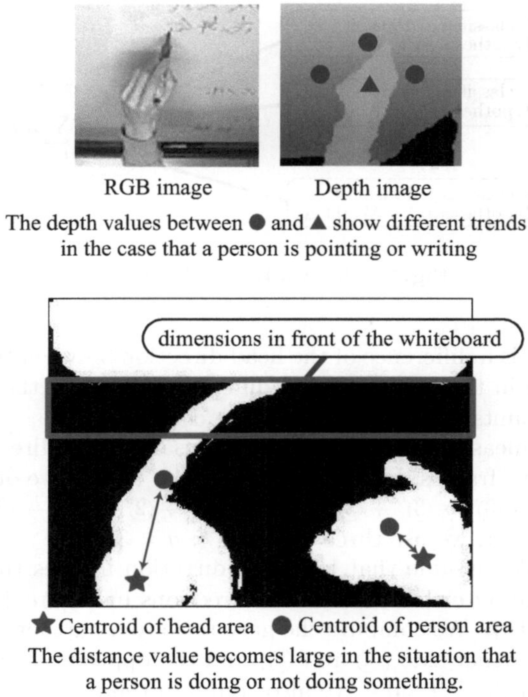

Fig. 3. The features for the operation identification

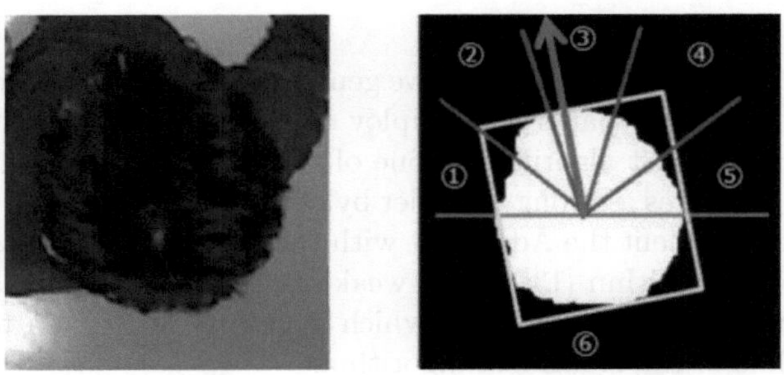

Fig. 4. Head direction

Center ratio. We compute a center ratio of two person. This ratio is which person stands to the center area in long time.

Average distance. We compute an average distance value between two persons from all image frames.

Operation ratio. We compute an operation ratio. Here the target operations are (1) pointing and (2) writing. This feature is the ratio of these two operations in all frames.

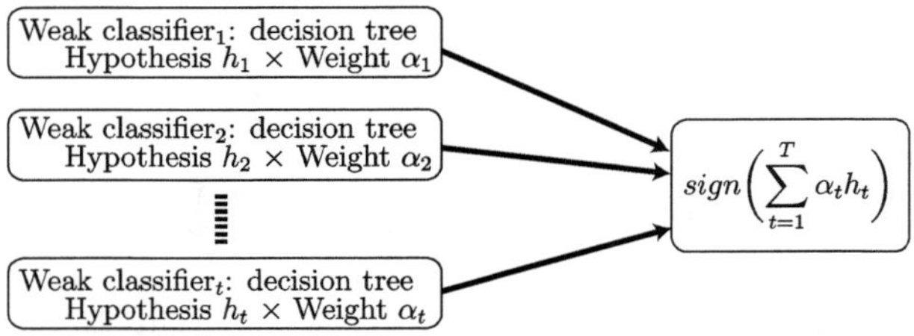

Fig. 5. The AdaBoost algorithm

Head turn. This is a difference of the head direction between two frames. If the head direction in the current frame differs from that in the previous frame, our method counts up the number of the changes.

Gaze ratio. We measure the frequency of the same head direction that occurs in n consecutive frames. For example, we set n to 3. If we obtain a direction sequence, ③ → ③ → ③ → ③ → ③ → ③ → ② → ② → ②, the frequency is 2. In this paper, we use three types of n; $n = 3, 5, 10$.

Eye direction. We assume that the head direction implies the gaze direction of eyes. We compute the ratios of six directions in Figure 4 in all frames.

Partner gaze ratio. We compute the gaze ratio to a partner. We compute the ratios of the partner side (④ and ⑤) and the opposite side (①, ② and ③) for a person in the left side. We also compute the ratios of the partner side (① and ②) and the opposite side (③, ④ and ⑤) for a person in the left side

3.3 Classifier

On the basis of the extracted features, we generate a classifier for the cooperation level estimation. In this paper, we employ the AdaBoost algorithm [1] as the classifier. The AdaBoost algorithm is one of the most famous machine learning techniques. It generates a strong classifier by combining some weak classifiers. In this paper, we implement the AdaBoost with the open source software Weka[17]. We use the C4.5 algorithm [13] as the weak classifiers[4]. C4.5 is also one of the famous machine learning techniques, which generates a decision tree. Figure 5 shows the outline of the AdaBoost algorithm.

In addition, we introduce another approach. We employ the multiple linear regression analysis with the stepwise method for the estimation.

4 Experiment

We evaluate our method with an annotated data set of pair work. In this section, we explain the experimental settings first. Then, we consider the relevancy between the cooperation level and some features. Finally, we discuss the results of two machine learning approaches.

[4] Actually, it is "48" in Weka.

4.1 Setting

We collected pair work with a whiteboard by using an overhead camera. The pair work consisted of two processes. First, each test subject wrote words related to a given topic to the whiteboard. In this experiment the topics were "Summer" and "Autumn". For example, the written words were "Fireworks", "Festival", "Summer holiday" and so on. Next, they classified the written words on the whiteboard into several categories in a subjective manner. For example, a pair categorized these three words, "Fireworks", "Festival", "Summer holiday", as "Summer event". The categorization process depended on the free discussion of each pair. Each operation time of two process, namely the listing of words and categorization, was three minutes, respectively.

The number of test subjects was 16 persons. We generated ten groups from them. Five groups have the acquaintance relationship and others were the first meet. The cooperation level of each pair was determined by one annotator[5]. The range of cooperation level was 1 (bad) to 5 (good).

We obtained 1800 frames from Kinect. We extracted 180 images from them with respect to each ten frame. We divided 180 images into 90 images as the anterior half and 90 images as the posterior half in each pair work.

4.2 Result

In this section, we discuss three main nonverbal aspects described in Section 3.1, namely standing location, operation ratio and head direction, in the experimental data first. Then, we discuss the accuracy of the AdaBoost and the reliability of the multiple linear regression approach.

Discussion on Location, Operation and Head Direction
Table 1 and Table 2 show the experimental data about acquaintance groups and first-meet groups, respectively. In the tables, CL denotes the cooperation level by one annotator. LocDev, CentRatio, AveDist and OpRatio are the location deviation, the center ratio, the average distance and the operation ratio in Section 3.2, respectively. "A" to "J" denote the group ID. "(M)" and "(F)" (F)" are male and female. "/" is the border of the anterior half and the posterior half about each feature. For example, the test subject A1 was male and the anterior half and the posterior half features of his location deviation were 31.97 and 37.80.

First, we discuss the location features. The LocDev became larger in the situation that the cooperation level was high, e.g., Group A and B. On the other hand, the cooperation levels about the groups or persons with the small LocDev

[5] Actually, we collected a questionnaire about the cooperation level of the pair work from the test subjects. However, there was a large difference between self-evaluation and cooperation level by the annotator. In other words, the self-evaluation differ from the actual cooperation level. We think that the reason was that each test subject was liable to pay mind to the partner in the pair work.

Table 1. Result of Group A to E

Acquaintance	CL	LocDev	CentRatio	AveDist	OpRatio
A1(M)	5/5	31.97/37.80	1.4/1.2	356.2/354.7	44.4/41.1
A2(M)	5/5	32.13/53.12	98.6/98.2		62.2/81.1
B1(M)	5/5	35.68/34.98	30.0/26.7	395.5/387.3	47.8/60.0
B2(F)	5/5	26.07/39.17	70.0/73.3		57.8/40.0
C1(M)	3/1	13.35/17.07	5.6/0.0	429.3/446.5	27.8/5.6
C2(F)	4/2	41.56/23.95	94.4/100.0		38.9/42.2
D1(M)	4/5	42.45/46.63	0.0/16.7	356.6/343.2	12.2/58.9
D2(M)	4/5	26.29/45.35	100/83.3		54.4/47.8
E1(F)	4/3	18.42/15.08	15.6/14.4	266.5/270.8	35.5/15.5
E2(F)	5/4	33.00/24.37	84.4/85.6		62.2/50.0

Table 2. Result of Group F to J

First-meet	CL	LocDev	CentRatio	AveDist	OpRatio
F1(M)	4/3	22.97/33.68	59.3/100.0	395.8/358.6	64.4/23.7
F2(F)	3/2	18.22/19.12	40.7/0.0		6.8/18.6
G1(M)	4/2	20.72/20.71	2.1/68.8	328.7/339.7	41.7/6.25
G2(F)	4/4	14.91/36.02	97.9/31.2		47.9/54.1
H1(M)	3/3	9.96/21.58	0.0/1,9	312.5/308.4	28.8/22.6
H2(M)	4/3	21.22/18.12	100.0/98.1		35.8/28.8
I1(M)	4/4	22.65/31.25	0.0/0.0	353.0/341.3	55.6/44.4
I2(F)	4/4	20.44/15.94	100/100.0		51.1/42.2
J1(M)	4/1	30.48/16.42	82.2/100.0	370.8/356.3	13.3/2.2
J2(F)	4/1	23.39/14.11	17.8/0.0		44.4/20.0

value, such as Group J and person C1, were low. In other words, persons with high cooperation level were active during the pair work in this experiment. This denotes that large and frequent motions of the head and body are generated in the case that the pair work proceeds smoothly. Furthermore, the CentRatio of persons with leadership potential tended to increase the value, such as A1 and C2. The reason was that such persons were enterprising and willing, e.g., writing on whiteboard. In addition, for a female pair, Group E, the AveDist had a tendency to become small.

Next, we discuss the operation ratio. The OpRatio became larger for the high cooperation level, such as Group A and B, and lower for the low cooperation level, such as Group J. This is the same tendency as the LocDev. In contrast, there was no significant difference between acquaintance and first-meet about each features.

Finally, we discuss the head direction. For the head direction features, there was no relativity about the cooperation level. The groups of the high cooperation level had a tendency to increase operation time about writing on whiteboard.

Table 3. Confusion matrix

	Estimated		
	Cooperative	Neither	Uncooperative
Cooperative	25	1	1
Neither	1	2	3
Uncooperative	1	2	4

Therefore, the persons of the group fixed their eyes to the same direction. On the other hand, the groups of the low cooperation level had a tendency to increase non-operation time because they stood by doing nothing. Therefore, the persons of the group also fixed their eyes to the same direction, namely the white board. As a result, there was no difference between the high and low cooperation groups.

Discussion on Classifiers
Next, we evaluated two machine learning approaches with our features. The annotated data consisted of 20 persons for 10 groups. We divided the data of 20 persons into two parts; the anterior half and the posterior half. Hence we obtained 40 instances as the experimental data. We generated a three-class problem from the five cooperation level. In other words, we integrated the cooperation level 4 and 5 to "Cooperative" and 1 and 2 to"Uncooperative". The cooperation level 3 was class "Neither"

We evaluated the AdaBoost with 20-fold cross validation. On the basis of a preliminary experiment, we used the location deviation and the operation ratio as the features of the AdaBoost. The accuracy rate was 77.5%. Table 3 shows the confusion matrix of the experimental result. Although the accuracy was relatively-good, the evaluation data was unbalance. Most instances were the class "Cooperative". To validate the effectiveness of our method, we need to acquire more pair work data. The reason why the class "Cooperative" became the great majority was that we applied pair work to the task. Pair work has a natural tendency to cooperate with each other because the group consists of only two persons. Extension to a multi-party task is the most important future work.

Finally, we evaluate the reliability of the multiple linear regression approach. For the task, we used the original five cooperation level in the data. We applied all nonverbal features to the regression. In the analysis, we select the optimal features from them by using the stepwise method. As a result, LocDev and OpRatio were selected as explanatory variables. The equation about the cooperation level (CL) is as follows:

$$CL = 1.256 + 0.0415 \times LocDev + 0.0353 \times OpRatio \tag{1}$$

The standardized partial regression coefficients of LocDev and OpRatio were 0.371 ($p < .005$) and 0.572 ($p < .001$), respectively. The adjusted R^2 was 0.649. The nonverbal features were effective because the $AdjR^2$ was very high. However, the data was annotated by one annotator. Therefore, we need to verify the annotated cooperation level by several annotators. This is also the important future work.

5 Conclusions

In this paper, we discussed the estimation of a cooperation level in pair work. The task was the cooperation work that take place in front of a whiteboard by two persons. This study leads to one useful task for education support systems and project based learning.

We explained several nonverbal features and analyzed real data by them. Some knowledge was acquired about relations between the cooperation level and the nonverbal information, such as the location deviation and operation ratio. On the other hand, the head direction features were not effective for the estimation of the cooperation level. We also applied the features to two machine learning approaches. The AdaBoost with the features produced 77.5% as the accuracy. The adjusted R^2 of the multiple linear regression was 0.649. However, it is important to investigate and incorporate new nonverbal features to our method for the improvement of the cooperation level estimation.

Future work includes (1) evaluation of the method with a large-scale dataset, (2) utilization of other nonverbal features, (3) improvement of the operation identification and the head direction estimation, (4) annotation of the cooperation level by several annotators and (5) extension to a multi-party task (cooperation by 3 or more persons).

References

1. Freund, Y., Schapire, R.E.: Experiments with a new boosting algorithm. In: Proceedings of International Conference on Machine Learning, pp. 148–156 (1996)
2. Grafsgaard, J.F., Boyer, K.E., Wiebe, E.N., Lester, J.C.: Analyzing posture and affect in task-oriented tutoring. In: Proceedings of FLAIRS Conference 2012 (2012)
3. Hmelo-Silver, C.E.: Problem-based learning: What and how do students learn? Educational Psychology Review 16, 235–266 (2004)
4. Jayagopi, D.B., Sanchez-Cortes, D., Otsuka, K., Yamato, J., Gatica-Perez, D.: Linking speaking and looking behavior patterns with group composition, perception, and performance. In: Proceedings of the International Conference on Multimodal Interaction (ICMI), Santa Monica, USA (2012)
5. Komatsu, K., Shimada, K., Endo, T.: Posture identification for evaluation of multi-party interaction. Technical Report of IEICE, HCS 112, 25–30 (2012)
6. Kouno, D., Shimada, K., Endo, T.: Person identification using top-view image with depth information. In: Proceedings of 13th ACIS International Conference on Software Engineering, Artificial Intelligence, Networking and Parallel/Distributed Computing (SNPD 2012), pp. 140–145 (2012)
7. Kumano, S., Otsuka, K., Matsuda, M., Yamato, J.: Understanding empathy/antipathy perceived by external observers based on behavioral coordination and response time. In: Proceedings of HCG Symposium (2012)

8. Mahmoud, M., Baltrušaitis, T., Robinson, P., Riek, L.D.: 3d corpus of spontaneous complex mental states. In: D'Mello, S., Graesser, A., Schuller, B., Martin, J.-C. (eds.) ACII 2011, Part I. LNCS, vol. 6974, pp. 205–214. Springer, Heidelberg (2011)
9. Mota, S., Picard, R.W.: Automated posture analysis for detecting learner's interest level. In: Proceedings of Workshop on Computer Vision and Pattern Recognition for Human-Computer Interaction, CVPR HCI (2003)
10. Nakamura, K., Kakusho, K., Murakami, M., Minoh, M.: Estimating learners' subjective impressions of the difficulty of course materials in e-learning environments. In: Proceedings of APRU 9th Distance Learning and Internet Canference 2008, pp. 199–206 (2008)
11. Nakatani, R., Kouno, D., Shimada, K., Endo, T.: A person identification method using a top-view head image from an overhead camera. Journal of Advanced Computational Intelligence and Intelligent Informatics 16(5), 696–703 (2012)
12. Omura, M., Shimada, K.: Estimation of subjective impressions of difficulty in quiz dialogue. Technical Report of IEICE NLC2013-2, 7–14 (2013)
13. Quinlan, J.R.: C4.5 Programs for Machine Learning. Morgan Kaufmann Publishers (1993)
14. Shimada, K., Kusumoto, A., Yokoyama, T., Endo, T.: Hot spot detection in multiparty conversation using laughing feature. Technical Report of IEICE NLC2012-7, 25–30 (2012)
15. Takashima, K., Fujita, K., Yokoyama, H., Itoh, Y., Kitamura, Y.: A study of nonverbal cues and subjective atmosphere in six person conversations. In: Proceedings of IEICE SIG-HCS, vol. 112, pp. 49–54 (2012)
16. Vargas, M.F.: Louder than Words: an Introduction to Nonverbal Communication. Iowa State Press (1986)
17. Witten, I.H., Frank, E., Hall, M.A.: Data Mining: Practical Machine Learning Tools and Techniques, 3rd edn. Morgan Kaufmann (2011)
18. Yamane, T., Nakamura, K., Ueda, M., Mukunoki, M., Minoh, M.: Detection of interaction between a lecturer and learners based on their actions. In: Proceedings of The Japanese Society for Artificial Intelligence, SIG-ALST(Advanced Learning Science and Technology) (2010)
19. Yokoyama, T., Shimada, K., Endo, T.: Hot spot detection in multi-pary conversation using linguistic and non-linguistic information. In: Proceedings of NLP 2012 (2012) (in Japanese)

Korean Word Search Interface for Wearable Computers Using a Wrist-Mounted Camera Device

Hyun Park[1,*], Hyo-Seok Shi[1], Heon-Hui Kim[1], Kwang-Hyun Park[2]

[1] Art and Robotics Institute, Kwangwoon University, Seoul, 139-701, Republic of Korea
[2] School of Robotics, Kwangwoon University, Seoul, 139-701, Republic of Korea
park1263@naver.com, {mni3004,heonhuikim}@gmail.com,
akaii@kw.ac.kr

Abstract. This paper presents a hand shape recognition method and its application to Korean word search interface for wearable computers. We first describe a hand shape recognition method composed of hand region segmentation and recognition. To enforce the recognition performance, a user adaptation process is also proposed in algorithmic details. A Korean word search system is then proposed, which is based on the recognition of Korean manual alphabets using a wrist-mounted camera device. The effectiveness of the proposed method is verified through several experiments for evaluating recognition performance.

Keywords: Wrist-mounted camera, Korean manual alphabet recognition, user adaptation, hand shape recognition.

1 Introduction

Gestures have been regarded to be one of the potentially promising manners for input interfaces in human-computer interaction (HCI) together with facial and vocal interfaces. By using those interfaces, a user is allowed to interact without any physical contact with a computer or machine, which is definitely distinguished from traditional mouse/keyboard/touch interfaces. For that reason, gesture-based interface systems have been studied widely and implemented in various forms: e.g., the Softremocon[1] for controlling home appliances and the SixthSense[2] for interacting with a wearable computer. One of the important aspects is that the gesture interface has been evolving toward a compact and modularized system for users to be able to feel convenient in practical application environments. In fact, such design considerations are more required in development of input interfaces for wearable computers.

Recently, several innovative systems have been invented in the field of wearable computer. As a representative wearable computer, the Google Glass enables a user to experience augmented reality through an optical head-mounted display. Even though the device has a number of potential abilities to reproduce interesting applications, it does not seem to provide sufficient modality to interact with its owner. A compact wearable device that provides gestural interface can be considered as an alternative to supplement the insufficient modality for interaction with a human being.

* Corresponding author.

S. Kim et al. (eds.), *Soft Computing in Intelligent Control*,
Advances in Intelligent Systems and Computing 272,
DOI: 10.1007/978-3-319-05570-1_10,

In this study, we have developed a compact wearable gesture interface that is configured in the form of a wrist-mounted camera. This device is carefully designed by a tradeoff between vision-based and globe-based recognition system for capturing hand postures. As a wrist-mounted camera configured vision system, the device is simply worn on human wrist and provides hand images in close proximity due to physical constraints. That is, our device can capture hand postures invariantly to hand pose as well as segment hand region better in visual processing actively by utilizing physical constraints formed from wrist-mounted configuration (See more details in [3]).

We are concerned in this paper with an application of our device to a word search interface that can be useful for other wearable computing devices. To do so, a Korean manual alphabet recognition system using a wrist-mounted camera device is first described. The proposed system is composed of a hand region segmentation algorithm which extracts a skin color region based on a set of adaptive histograms, and a user adaptation algorithm that leads a user-independent model (IM) gradually to a certain user-dependent model (DM). The word search interface is then implemented based on the recognition of Korean manual alphabets. This paper evaluates overall recognition performance with respect to Korean manual alphabets, and shows some examples of word search results.

2 Korean Manual Alphabet Recognition System

2.1 Hand Region Segmentation

Because every image captured from the wrist-mounted camera always contains a user's hand, we utilize the rectangular region in the bottom of the image as a reference mask to extract skin color information of hand region. Hand region segmentation algorithm distinguishes possible hand regions from background by using two HSV histograms for skin color and non-skin color, which are updated for each image frame by Bayesian rules [4-13]. Fig.1 shows the initialization process of a HSV histogram. Because the reference mask also contains both hand and background, the segmentation process first divides the mask region into two regions filled with skin color and non-skin color. For skin-colored regions, morphological operation is adopted to remove image noises. Two HSV histograms for skin and non-skin color are then constructed after being converted from the original RGB color for the mask image.

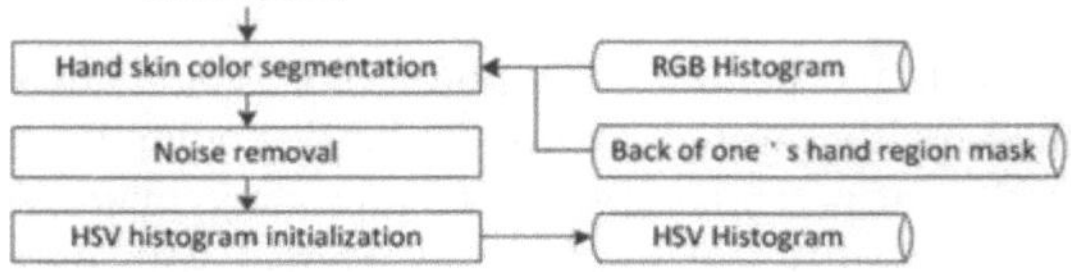

Fig. 1. HSV Histogram initialization

Fig.2 depicts the adaptation process of the HSV histograms. In the adaptation stage, preprocessing algorithms are the same as those of the initialization stage, but histogram update is carried out twice. The HSV histograms are first updated using the pixel information of whole image regions, and then updated once more using the pixel information of only the hand regions extracted from the skin color segmentation process.

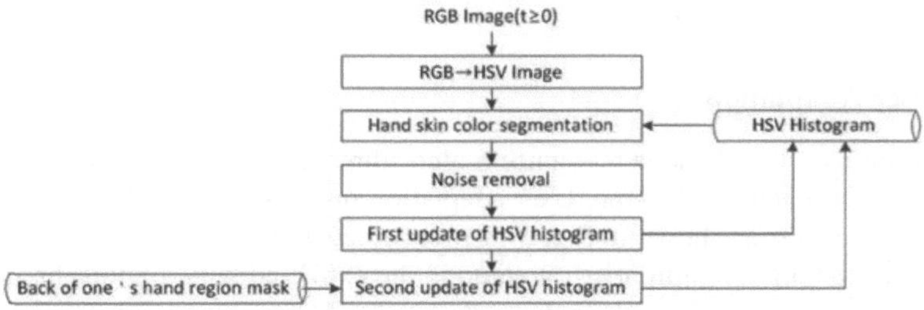

Fig. 2. HSV Histogram adaptation process

For being used for the next step, the HSV histogram is updated from the last histogram and the new histogram by Eq. (1).

$$H(l,m,n) \leftarrow \alpha H^{new}(l,m,n) + (1-\alpha)H(l,m,n), \quad l,m,n \in \{0,1,\cdots.31\}, 0<\alpha<1 \tag{1}$$

where l, m, and n denote the indicates of 3-demensional bin for a HSV histogram. The parameter α stands for a weight parameter in range of zero to one, and is set to be 0.6 for skin color space and 0.8 non-skin color spaces in this paper.

2.2 Hand Shape Recognition

2.2.1 User Model

For hand shape recognition, we define two types of user models, i.e. user-dependent model (DM) and user-independent model (IM). Both models consist of a binary image $B^n_{i,j}$ whose value of pixel coordinate (x, y) is 1 for hand region and 0 for background. Let *i, j* be the indices of hand shape and user such that $i \in \{1,2,\cdots,I\}$ and $j \in \{1,2,\cdots,J\}$ where the number of hand shapes and users are denoted by *I* and *J*, respectively. Then, the user-dependent model for the jth hand shape of the ith user is represented by $DM_{i,j}$, and computed for each pixel coordinate (x, y) as Eq. (2).

$$DM_{i,j}(x,y) = \frac{\sum_{n=1}^{N_i} B^n_{i,j}(x,y)}{N_i} \tag{2}$$

For the jth hand shape, user-independent model denoted by $IM_j(x,y)$ is defined to be marginalization with respect to all the users as Eq. (3).

$$IM_j(x,y) = \frac{\sum_{i=1}^{I}\sum_{n=1}^{N_i} B_{i,j}^n(x,y)}{\sum_{i=1}^{I} N_i} \quad (3)$$

2.3 User Adaptation

For a certain user, hand shape recognition algorithm starts with a set of user independent models for each hand shape. User adaptation process updates the IMs separately for each hand shape as shown in Fig. 3. It first detects the index of a hand shape model which has the highest similarity rate among the extracted hand regions based on the user-independent model. The algorithmic details for determining the hand shape index are described in Algorithm 1.

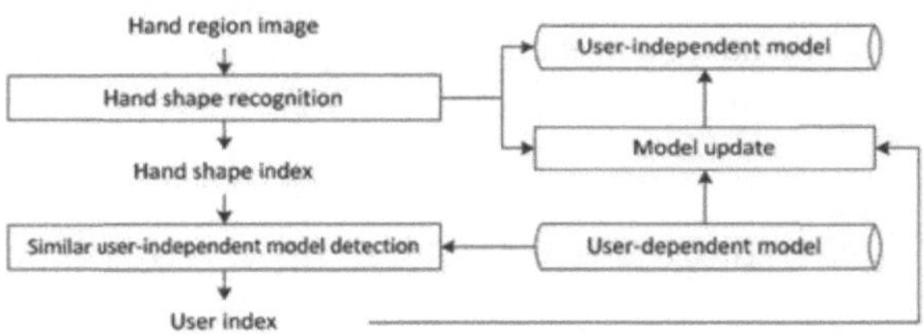

Fig. 3. User adaptation process

Algorithm 1. Determination of hand shape index

IM_j = User-independent model of hand shape index j
S_j = Similarity rate of hand shape index j
x, y = pixel coordinate in image
$B(x,y)$ = Binary image of segmented hand region
for all j
    $S_j = 0$
    for all (x,y)
        if $B(x,y) == 1$ then $S_j = S_j + IM_j(x,y) - 0.5$
        else $S_j = S_j - IM_j(x,y) + 0.5$
$j_{\max} = \arg\max S_j$

In Algorithm 1, the index j_{max} can be regarded as the recognition result of hand shape which has the highest similarity rate between user-dependent models and the current hand shape. The adaptation process finally determines the index of the most possible user by a winner-take-all strategy.

After this process, the user-independent model is updated with its binary image for the recognized hand model j_{max} and then it updates the user-independent model with the rest of hand shape model which has the highest similarity rate with the current user. User-independent model for hand shape index j_{max} is updated by Eq. (4).

$$IM_{j_{max}}(x,y) \leftarrow \gamma B(x,y) + (1-\gamma) IM_{j_{max}}(x,y) \quad (4)$$

Also, user-independent models for the remaining hand shapes are updated for every index ($j \neq j_{max}$) by Eq. (5).

$$IM_{j}(x,y) \leftarrow \gamma DM_{i_{max},j}(x,y) + (1-\gamma) IM_{j}(x,y) \quad (5)$$

Here, the parameter γ represents a weight factor, and is set to be 0.005 in this paper.

3 Word Search Interface for Wearable Computers

Based on the hand shape recognition system, we have developed a Korean word search interface as an application example of the proposed method. This interface system can be used for wearable computing devices, and provide a useful function for searching a list of words by a wrist-mounted camera device. In this paper, the system can recognize 18 hand shapes that consist of 16 Korean manual alphabets (Fig. 4) and two of additional hand signs (Fig. 5).

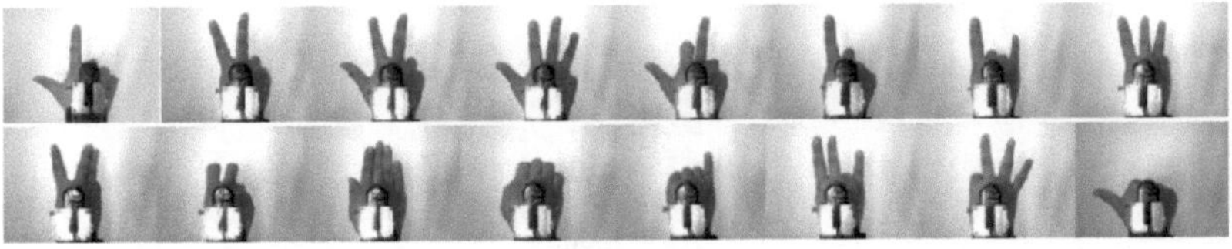

Fig. 4. Korean manual alphabet

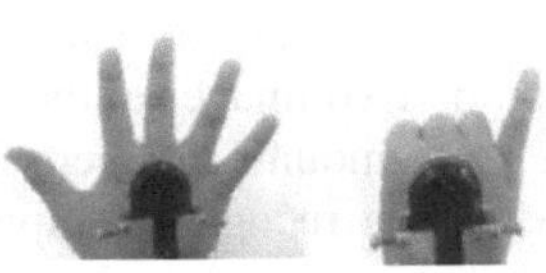

Fig. 5. Hand signs for indicating "begin" and "search" commands

Fig. 6 illustrates how a set of control commands is inputted into the search system as an example. The five-fingered hand corresponds to the control command for "begin recognition" while the little-fingered hand indicates "search word". In Fig. 6, the hand shapes in (b) and (d) represents the Korean manual alphabets "ㄱ","ㄱ", respectively. Thus, the example as shown in Fig. 6 corresponds to the control commands "search a Korean word for a given Korean consonant sequence of { ㄱ, ㄱ}".

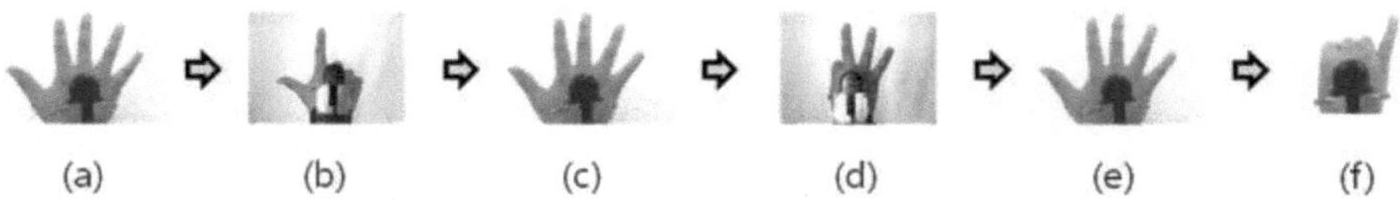

Fig. 6. Application execution order(Inputted Korean manual alphabet "ㄱ","ㄱ")

4 Experiment Results

We tested the recognition performance of the proposed system for 16 Korean manual alphabets as presented in Fig. 4. Fig. 7 shows an adaptation trend from a user-independent model, which is gradually converging to a user-dependent model. By using user-independent model, the recognition system could recognize the hand shapes with 63.17% of recognition rate in this experiment. As time goes on after 1500 frame, the recognition performance could be increased with 88.17% of recognition rate.

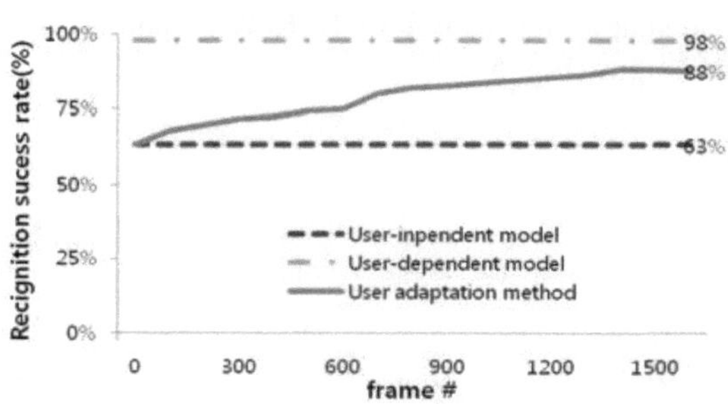

Fig. 7. Change of recognition success rate

Fig. 8 depicts an example result of Korean word search system for a given sequential gesticulation as shown in Fig. 6. Two images at the bottom of the figure represent an input image captured by the wrist-mounted camera device and the result of hand region segmentation, respectively. The recognized alphabets are displayed together with a word list for a given alphabet sequence at the top of the figure. Note that the word database used in this experiment is constructed only by the universities placed in Seoul, Korea.

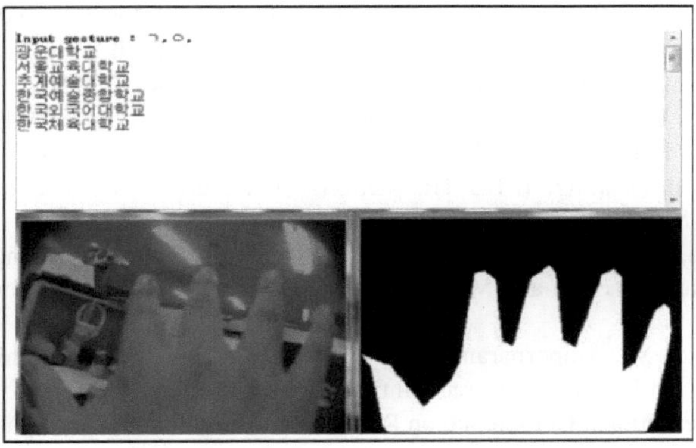

Fig. 8. Interface of application

5 Conclusion

This paper has dealt with a Korean manual alphabet recognition system and its application to Korean word search interface for wearable computers. We have first proposed a hand shape recognition method that consists of hand region segmentation and hand region recognition algorithms. For being better utilized in practical applications, the proposed system was designed to contain a user adaptation process based on user-independent models and user-dependent models. As an application of the proposed method to wearable computing devices, we have proposed a Korean word search interface system. In this paper, the proposed system was tested through several experiments for evaluating recognition performance, and we found that the user adaptation process could contribute to enhancing the recognition performance up to 25% increase of recognition rate.

References

1. Do, J.-H., Kim, J.-B., Park, K.-H., Bang, W.-C., Bien, Z.Z.: Soft Remote Control System using Hand Pointing Gestures. International Journal of Human-Friendly Welfare Robotic Systems (HWRS) 3(1), 27–30 (2002)
2. Mistry, P., Maes, P.: SixthSense – A Wearable Gestural Interface. In: Proceedings of SIGGRAPH Asia 2009, Yokohama, Japan (2009)
3. Park, H., Shi, H.-S., Kim, H.-H., Park, K.-H.: A User Adaptation Method for Hand Shape Recognition Using Wrist-Mounted Camera. The Journal of the Korea Institute of Electronic Communication Sciences 8(6), 805–814 (2013)
4. Hamada, Y., Shimada, N., Shirai, Y.: Hand shape estimation using image transition network. In: Proceedings of Workshop on Human Motion, pp. 161–166 (2000)
5. Ong, E.-J., Bowden, R.: A boosted classifier tree for hand shape detection. In: Proceedings of Sixth IEEE International Conference on Automatic Face and Gesture Recognition, pp. 889–894 (2004)

6. Athitsos, V., Sclaroff, S.: An appearance-based framework for 3D hand shape classification and camera viewpoint estimation. In: Proceedings of Fifth IEEE International Conference on Automatic Face and Gesture Recognition, pp. 40–45 (2002)
7. Quek, F.K.H.: Unencumbered gestural interaction. IEEE MultiMedia 3(4), 36–47 (1996)
8. Ong, S.C.W., Ranganath, S.: Automatic sign language analysis: a survey and the future beyond lexical meaning. IEEE Transactions on Pattern Analysis and Machine Intelligence 27(6), 873–891 (2005)
9. Yu, Z., Xilin, C., Debin, Z., Hongxun, Y., Wen, G.: Adaptive sign language recognition with exemplar extraction and MAP/IVFS. IEEE Signal Processing Letters 17(3), 297–300 (2010)
10. Hasanuzzaman, M., Ampornaramveth, V., Zhang, T., Bhuiyan, M.A., Shirai, Y., Ueno, H.: Real-time vision-based gesture recognition for human robot interaction. In: Proceeding of the IEEE International Conference on Robotics and Biomimetices, August 22-26, pp. 413–418 (2004)
11. Park, K.H.: Betterment of Mobile Sign Language Recognition System. IEEK System and Control 43(1), 1–10 (2006)
12. Jones, M.J., Rehg, J.M.: Statistical color models with application to skin detection. In: Proceedings of the IEEE Computer Vision and Pattern Recognition, pp. 274–280 (1999)
13. Sigal, L., Sclaroff, S., Athitsos, V.: Skin color-based video segmentation under time-varying illumination. IEEE Transactions on Pattern Analysis and Machine Intelligence 26(7), 862–877 (2004)

Author Index

MIX
Papier aus verantwortungsvollen Quellen
Paper from responsible sources
FSC® C105338

If you have any concerns about our products,
you can contact us on
ProductSafety@springernature.com

In case Publisher is established outside the EU,
the EU authorized representative is:
Springer Nature Customer Service Center GmbH
Europaplatz 3, 69115 Heidelberg, Germany

Printed by Libri Plureos GmbH
in Hamburg, Germany